Instrumental Methods of Analysis

Instrumental Methods of Analysis

Dr. C.N. Nalini, M.Pharm., Ph.D,
Professor,
Head of the Department of Pharmaceutical Analysis
C.L. Baid Metha College of Pharmacy,
Chennai, Tamil Nadu.

PharmaMed Press

An imprint of BSP Books Pvt. Ltd.
4-4-309/316, Giriraj Lane,
Sultan Bazar, Hyderabad - 500 095.

Instrumental Methods of Analysis
by **Dr. C.N. Nalini**

Disclaimer: The authors and publishers have taken due care to provide authentic, reliable and up-to-date information related to the subject. However, neither the authors nor the publisher shall be responsible for any liability for any damage caused because of the use of this book. The respective users must check the accuracy from other sources too.

Published by

PharmaMed Press

An imprint of BSP Books Pvt. Ltd.

4-4-309/316, Giriraj Lane, Sultan Bazar, Hyderabad - 500 095.

Phone: 040-23445688; Fax: 91+40-23445611

E-mail: info@pharmamedpress.net

www.bspbooks.net/www.pharmamedpress.net

ISBN: 978-93-91910-95-2 (Hardback)

Preface

In recent times, we come across various powerful analytical techniques for obtaining qualitative and quantitative information about the composition and structural elucidation of different kinds of compounds. These instrumental techniques are applicable to various fields including pharmacy, chemistry, biochemistry, physics, geology, life sciences, forensic and environmental sciences. A clear understanding of the instrumental methods and their applications helps in solving the analytical challenges in these fields. The making of appropriate decisions and efficient usage of the analytical instruments can be done only if the concepts behind the techniques are known.

The objective of the book is to impart fundamental knowledge on the principles and instrumentation of some of the most commonly used spectroscopic and chromatographic techniques. This book caters to the need of the students and analysts in various fields for a thorough understanding of the construction of various analytical instruments and their principles. A detailed study of the book will make the readers proficient in spectroscopic and chromatographic techniques that are covered in the book, their applications including limitations.

The book is divided into two sections: The first section covers the spectroscopic techniques including UV Visible spectroscopy, Fluorimetry, IR spectroscopy, Flame Photometry, Atomic Absorption Spectroscopy and Nepheloturbidimetry. All these topics include a detailed description of the concepts involved, instrumentation and applications of different techniques with illustrations.

The second section describes the different chromatographic techniques which include an introduction to chromatography. The different techniques covered under chromatography are Adsorption and Partition column chromatography, Thin layer chromatography, Paper chromatography, Electrophoresis, Gas chromatography, High performance liquid chromatography (HPLC), Ion exchange chromatography, Gel chromatography and Affinity chromatography. This section gives a detailed note on various terminologies used in chromatography along with theoretical knowledge on the methodology, stationary and mobile phases along with their applications and limitations.

It gives me immense pleasure to express my gratitude to the management of C.L.Baid Metha College of Pharmacy for the support in pursuing this book. I convey my sincere thanks to Dr. Grace Rathnam, Dr. N. Ramalakshmi and Dr. S. Amuthalakshmi for their constant encouragement. I would like to place my acknowledgment and gratitude to Ms. Poorna Basuri for going through the entire script and aiding me with her suggestions. I also thank Ms. Basuri for designing the cover page for the book. My sincere gratitude to my family, my spouse Mr. K. Dinakaran, my daughters D. Varshini and D. Nithyashree for their understanding, patience and for giving me confidence.

Suggestions and criticisms for the improvement of the book shall be highly appreciated.

-Author

Contents

About the Author

The author, C N Nalini, is currently serving as Professor and Head of the Department of Pharmaceutical Analysis in C.L.Baid Metha College of Pharmacy, Chennai, Tamil Nadu. She has a teaching experience of 25 years along with two years of industrial experience as an analytical chemist. She has also served as a Pharmacist in Christian Fellowship Hospital, Oddanchatram and as a Senior Assistant Pharmacist for one year in Christian Medical College Hospital, Vellore.

She has completed her D.Pharm in 1985, B.Pharm in the year 1992, and M.Pharm in 1997. She was awarded a Ph.D. by the Tamil Nadu Dr. M.G.R. Medical University in the year 2010. She has been awarded the best teacher award from the Tamil Nadu Dr. M.G.R. Medical University in the year 2011.

She has held various positions like Member of Board of Studies, Research proposal screening committee, Screening committee for Research Grant, Inspection Committee, the expert Committee for Pharm D, B.Pharm and M.Pharm, Faculty at the TN Dr. M.G.R Medical University. She is also a member of the doctoral committee for research in various universities including TNMGRMU. She is a guide and co-guide for many doctoral students. She has guided more than 50 postgraduate students and many undergraduate students. She is a Question paper setter, valuator, practical examiner, and UG/PG dissertation adjudicator in the TN Dr. MGR Medical University and other universities including Anna University Chennai, Annamalai University, SRMCP, SRM, Mother Teresa, and Chettinad University.

She has more than 60 publications to her credit in peer-reviewed National and International journals. Her areas of publication include Pharmaceutical Analysis and Pharmaceutical Chemistry.

SECTION – I: Spectroscopy

1. UV Visible Spectroscopy
2. Fluorimetry
3. IR Spectroscopy
4. Flame Photometry
5. Atomic Absorption Spectroscopy
6. Nepheloturbidimetry

CHAPTER 1

UV Visible Spectroscopy

INTRODUCTION

Spectroscopy is the branch of science that deals with studying the **interaction of matter with electromagnetic radiation**. Due to this interaction, energy is either absorbed or emitted by the matter in discrete amounts called quanta. The measurement of this absorbed or emitted radiation by the matter forms the basis of spectroscopic techniques. Spectroscopic techniques are helpful in the study of atomic and molecular structures. Spectroscopy is one of the most powerful tools available for the qualitative and quantitative analysis of different samples.

Spectroscopy can be studied under two major headings

1. **Atomic spectroscopy** – deals with the interaction of electromagnetic radiation with atoms in the ground energy level (lowest energy).

   E.g., Atomic absorption spectroscopy, Atomic emission spectroscopy, Flame emission spectroscopy

2. **Molecular spectroscopy** – deals with the interaction of electromagnetic radiation with molecules.

   E.g., IR spectroscopy, Raman spectroscopy.

Spectroscopy can also be classified based on the measured mode of interaction of matter with electromagnetic radiation, which is given in the Table 1.1.

Table 1.1 Instrumental Methods based on measurement of property

Absorption of radiation	Emission of radiation	Scattering
Absorption spectroscopy: X-ray, UV, IR, Colorimetry, Atomic absorption, Nuclear magnetic resonance, and electron spin resonance spectroscopy	**Emission spectroscopy:** Flame emission photometry, Fluorescence spectroscopy	Turbidimetry, Nephelometry, Raman spectroscopy.

PROPERTIES OF ELECTROMAGNETIC RADIATION

Electromagnetic radiation is an energy form that travels with enormous velocity in space. Electromagnetic radiation (EMR) possesses both wave and particle nature. These properties are a distinct feature of EMR, and they are inseparable.

Wave properties of Electromagnetic radiation

Electromagnetic radiation has an electrical component and a magnetic component oscillating in perpendicular planes and also perpendicular to the direction of propagation. The wave nature of EMR shows properties like refraction, reflection, constructive and destructive interferences.

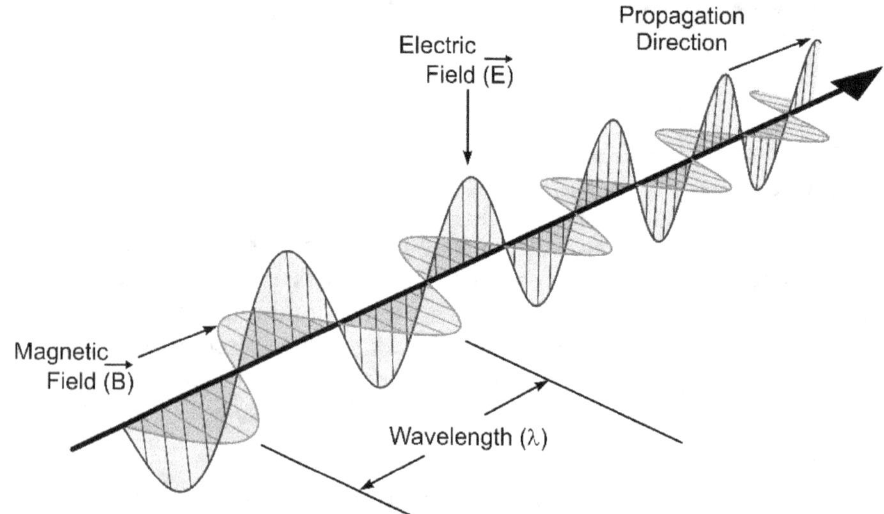

Fig. 1.1 Electromagnetic wave with its components.

The electromagnetic wave parameters are

(a) **Wave length, λ:** Is the **distance between two successive maxima or minima in an electromagnetic wave**. It is denoted by the Greek letter lambda. Units of wavelength are meters (m), centimeter (cm), millimeter (mm), micrometers (μm), nanometers (nm) and angstrom (Å).

$$1Å = 10^{-8}\ cm = 0.1nm = 10^{-10}m$$

A beam of radiation with only one wavelength is called **monochromatic radiation,** and the beam consisting of several wavelengths is called **polychromatic or heterochromatic radiation.**

(b) **Frequency, υ:** Is the **number of wavelength units passing through a given point in unit time.** It is denoted by the Greek letter υ. Unit is cycles per second or Hertz.

(c) **Wave number $\bar{\upsilon}$:** Frequency is usually a large number making its use more difficult in regular practice. To overcome this problem, frequency is usually expressed as wave number. **The wave number is the number of waves in vacuum.** The symbol is $\bar{\upsilon}$. Unit for wave number is cm^{-1}

$$\bar{\upsilon} = 1/\lambda$$

Particle properties of electromagnetic radiation

When electromagnetic radiation hits a material, it emits electrons called photoelectrons. This **photoelectric effect** is due to the particle properties of EMR. To explain this property, it is assumed that EMR consists of a stream of discrete packets (particles) of energy called photons or quanta. The energy of the photon is proportional to the frequency of the radiation and is given by the relationship

$$E = h\upsilon, \text{ where,}$$

E is the energy of a photon in ergs, υ is the frequency of the EMR in cycles per second, and h is called Planck's constant (6.624×10^{-34} joules/second).

EMR of longer wavelength (low frequency) has lower energy than the shorter wavelength (high frequency).

Electromagnetic spectrum

The entire range of EMR is called as EMR spectrum.

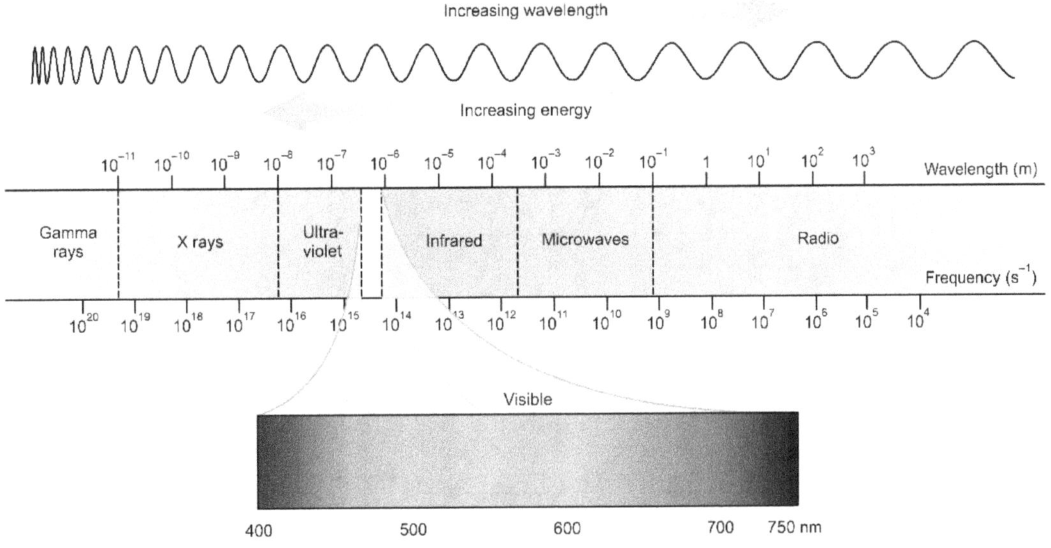

Fig. 1.2 Electromagnetic spectrum showing different regions of EMR.

The various spectral regions are

γ(Gamma) ray region – lies between 0.02 to 1 angstrom. These are the shortest waves.

X-ray region –This region lies between 1 to 10 angstrom. These rays cause core electron excitation.

Visible and Ultra violet region – These rays cause valence electron excitation. This region consists of three parts

   Vacuum UV : 1 - 180 nm

   Ultraviolet : 180 - 400 nm

   Visible   : 400 - 800 nm

Infrared region – This region is subdivided into three regions

   Near IR : 0.7 - 2.5 μm

   Mid IR   : 2.5 - 15 μm

   Far IR   : 15 - 200 μm

Microwave region – 0.1 mm – 1 cm

Radiofrequency region – 100m – 1 cm

INTERACTION OF ELECTROMAGNETIC RADIATION WITH MATTER

When electromagnetic radiation passes through matter, a variety of events may occur. Some of them are

1. **Absorption of radiation:** If the radiation of appropriate energy is used, they may be absorbed by the matter resulting in electronic, vibrational, and rotational changes or a combination of these changes. After absorption, the molecules become excited and go to a high energy level (low stability). They lose energy in the form of heat or re-emit EMR to come back to the ground state (high stability).

   The total energy of the molecule at the ground state is given as

   $$E_0 = E_{electronic} + E_{vibrational} + E_{rotational}$$

   The schematic energy level diagram for a simple diatomic molecule is shown in the figure.

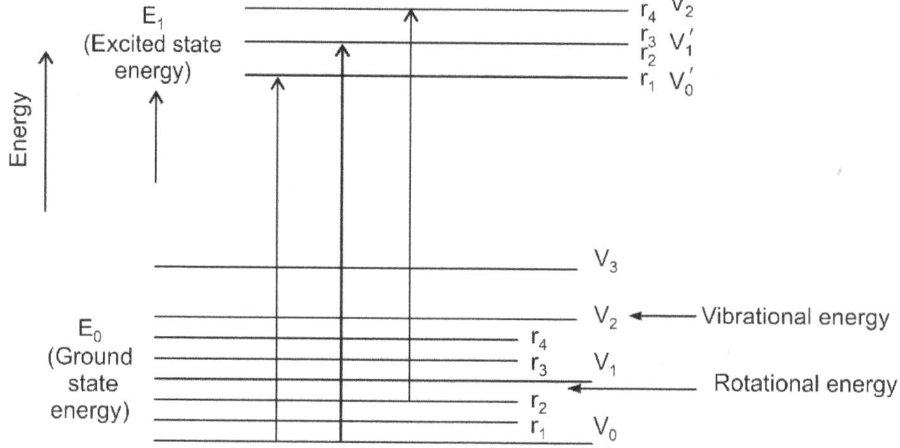

Fig. 1.3 Energy level diagram of a molecule showing vibrational and rotational energy level.

   An electronic transition due to absorption of light may include vibrational or rotational changes.

2. **Scattering:** Sometimes, the radiation is not absorbed completely by the matter, and a portion of it may undergo scattering or reflection

3. **Emission:** In some cases, the molecules become excited after absorbing the radiation, and they lose energy by re-emitting radiation either instantaneously or with a time delay.

Franck–Condon principle

Franck–Condon principle gives the interaction between the electronic and vibrational motions. It states that the timescale of electronic transitions are so rapid when compared with the nuclear motion and the mass of the nuclei are heavy compared with electrons. An electronic transition is therefore considered to be a vertical transition. The light is absorbed in femtoseconds to nanoseconds, and within this timescale, electrons move, but the nuclei cannot. The heavier atomic nuclei cannot adjust during the absorption but readjust after the absorption process, creating vibrations. Simultaneous occurrence of electronic and vibrational transitions is called vibronic transitions, which give rise to the vibrational structure of the electronic bands.

The quantum mechanical formulation of this principle is that the **intensity of a vibronic transition is proportional to the square of the overlap integral between the vibrational wave functions of the two states that are involved in the transition.**

The Figure illustrates the Franck–Condon principle for vibronic transitions in a molecule. Potential energy functions in both the ground and excited electronic states are shown. In the low temperature, the molecule in the $v = 0$ vibrational level of the ground electronic state absorbs a photon of the necessary energy, makes a transition to the excited electronic state. The electron configuration of the excited state results in a shift of the equilibrium position of the nuclei of the molecule. In the figure, this shift in nuclear coordinates between the ground and the first excited state is labeled as q_{01}. In a simple case of a diatomic molecule, the nuclear coordinate axis refers to the internuclear separation. The vibronic transition is indicated by a vertical arrow, assuming constant nuclear coordinates during the transition. In the electronic excited state, molecules quickly relax to the lowest vibrational level of the lowest electronic excitation state (Kasha's rule). From there, they decay to the electronic ground state through photon emission. The Franck–Condon principle is applied equally to absorption as well as fluorescence.

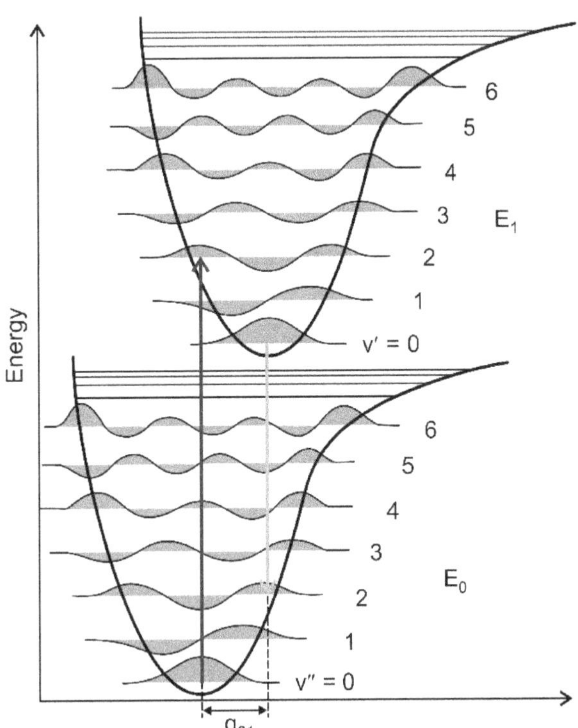

Fig. 1.4 Franck-condon principle for vibronic transitions in a molecule.

1.1 VISIBLE SPECTROPHOTOMETRY AND COLORIMETRY

UV-Visible Spectroscopy is one of the important absorption spectroscopic techniques, and it is most commonly employed for quantitative and qualitative analysis. It is also playing a role in the structural elucidation of organic compounds.

The wavelength range of visible light is 800-400 nm (8000-4000 Angstrom). This range of visible radiation is used for colorimetric analysis. Determination of concentration of coloured compounds in solution is called colorimetry. A colorimeter is a device used to test the concentration of a solution by measuring its absorbance at a specific wave length of light.

Coloured substances absorb in the visible region, and the colorless substances absorb in the UV region. The amount of light absorbed differs with the wavelength, and a plot of absorbance vs. wavelength is called an absorption spectrum. **In the absorption spectrum, the wavelength at which maximum absorption occurs is called λ_{max}.** This λ_{max} is characteristic or unique for every substance and is used to identify the substances (qualitative analysis). λ_{max} is **independent of concentration, meaning that the λ_{max} will not change** even if the concentration is changed.

A calibration graph is a plot of concentration vs. absorbance. This graph helps to determine the concentration/amount of the drug present in the given sample solution (quantitative analysis).

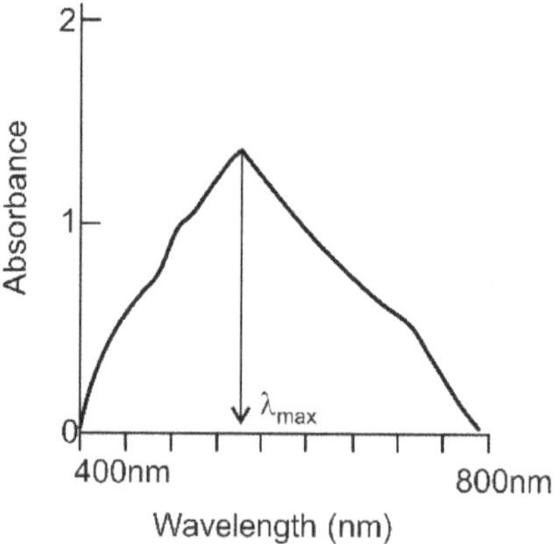

Fig. 1.5 Absorption spectrum.

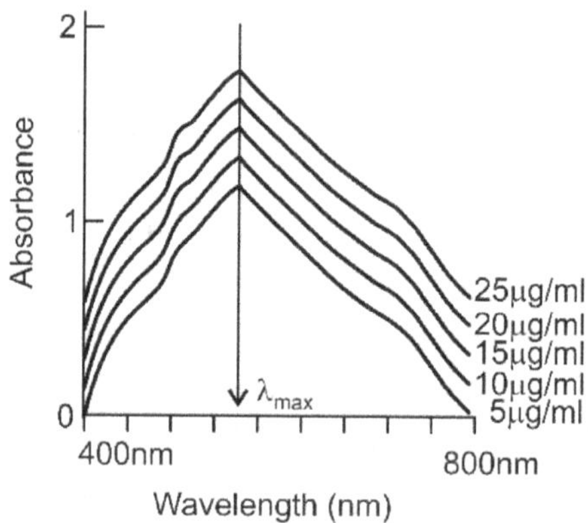

Fig. 1.6 Absorption spectrum of a substance in different concentrations.

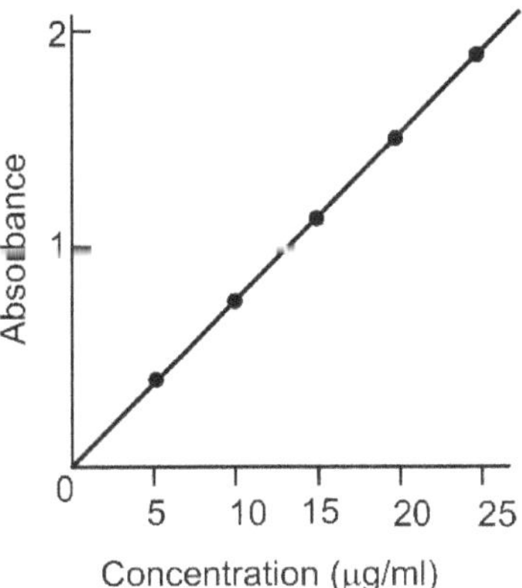

Fig. 1.7 Calibration curve.

THEORY OF SPECTROPHOTOMETRY & COLORIMETRY (ABSORPTION LAWS)

When light is incident upon a homogenous medium, a part of the incident light is reflected, a part is absorbed, and the remainder is transmitted so

$$I_0 = I_a + I_t + I_r \qquad .....(1.1)$$

Where I_0 – incident light; I_a – absorbed light; I_t – transmitted light and I_r – is reflected light. The value of I_r is minimal (about 4%), and it is eliminated for air-glass interfaces if a comparison cell is used. So equation (1.1) becomes as

$$I_0 = I_a + I_t \qquad .....(1.2)$$

The two laws governing the absorption of light are:

1. **Beer's law** (concentration-dependent)
2. **Lambert's law** (Thickness/path length dependent absorbance)

Lambert's law states that when a beam of monochromatic light is allowed to pass through a transparent medium, **the rate of decrease of intensity with the thickness of the medium is directly proportional to the intensity of incident light.**

$$-\frac{dI}{dt} \alpha\, I$$

$$-\frac{dI}{dt} = k\,I \qquad .....(1.3)$$

$$-\frac{dI}{I} = kdt \qquad \text{(rearranging the equation)}$$

On integration

$$-\ln I_t = kt + b \quad \text{(ln – natural logarithm)} \qquad .....(1.4)$$

where b is the integration constant

When thickness is zero, there is no absorbance. So $I_t = I_0$

Substituting this in equation 1.4

$$- \ln I_0 = k \times 0 + b$$

$$- \ln I_0 = b$$

Substituting this value of b in equation 1.4

$$- \ln I_t = kt - \ln I_0$$

On rearranging

$$\ln I_0 - \ln I_t = kt$$

$$\ln \frac{I_0}{I_t} = kt$$

Removing natural logarithm

$$I_0/I_t = e^{kt}$$

Taking inverse on both sides

$$I_t/I_0 = e^{-kt}$$

$$\mathbf{I_t = I_0.e^{-kt}} \qquad \qquad .....(1.5)$$

This equation is called as Lambert's law where I_0 is intensity of incident radiation, I_t is transmitted radiation and k is a constant.

Lambert's law shows the logarithmic relationship between transmittance and the optical path length

Beer's law: Beer observed a similar relationship between transmittance and the concentration of a solution. Beer's law states that the **intensity of a beam of monochromatic light decreases exponentially with an increase in the concentration of absorbing species arithmetically.**

$$- \frac{dI}{dc} \, \alpha I$$

$$- \frac{dI}{dc} = k'I \qquad \qquad .....(1.6)$$

Rearranging the equation

$$- \frac{dI}{I} = k'dc$$

On integration

$$- \ln I_t = k'c + b \ (\ln - \text{natural logarithm}) \qquad \qquad .....(1.7)$$

where b is the integration constant

When concentration is zero, there is no absorbance. So $I_t = I_0$

Substituting this in equation 1.7

$$- \ln I_0 = k' \times 0 + b$$

$$- \ln I_0 = b$$

Substituting this value of b in equation 1.7

$$-\ln I_t = k'c - \ln I_0$$

On rearranging

$$\ln I_0 - \ln I_t = k'c$$

$$\ln \frac{I_0}{I_t} = k'c$$

Removing natural logarithm

$$I_0/I_t = e^{k'c}$$

Taking inverse on both sides

$$I_t/I_0 = e^{-k'c}$$

$$\mathbf{I_t = I_0.e^{-k'c}} \qquad\qquad .....(1.8)$$

This equation is called as Beer's law equation where I_0 is intensity of incident radiation, I_t is intensity of transmitted radiation and k is a constant.

On combining equations 1.5 and 1.8, we get

$$\mathbf{I_t = I_0.e^{-Kct}}$$

on changing from natural logarithms to base 10,

$$\mathbf{I_t = I_0.10^{-Kct}} \text{ where } K = k \times 0.4343$$

Rearranging terms

$$I_t/I_0 = 10^{-Kct}$$

Taking inverse on both sides

$$I_0/I_t = 10^{Kct}$$

Taking log on both sides

$$\log I_0/I_t = Kct \qquad\qquad .....(1.9)$$

Transmittance $T = I_t/I_0$, Absorbance $A = \log 1/T$

Substituting for T in absorbance A, the equation becomes

$$A = \log I_0/I_t \qquad\qquad .....(1.10)$$

From equations 1.9 and 1.10

$$A = Kct.$$

If a is used instead of K, then

$$A = act$$

$$\mathbf{\log I_0/I_t = act}$$

is the mathematical expression of Beer Lambert's law,

Where A= absorbance, a = absorption coefficient or absorptivity, c= concentration of the substance and t = path length in cm.

Absorptivity reflects the sensitivity of the procedure. Sensitivity increases with an increase in absorptivity.

If the concentration is expressed in moles/liter, then a is replaced with ε. Then the equation becomes

$$A = \varepsilon c t \qquad \qquad \qquad .....(1.11)$$

where

A = absorbance, ε = molar absorption coefficient or Molar absorptivity, c = concentration of the substance in mol/lit, t = path length in cm.

A = ε, when the concentration is one mole per liter and the path length is one centimeter.

Thus **Molar absorption coefficient** is defined as the absorption of a one molar solution when the path length is one cm. The molar absorption coefficient is otherwise called molar absorptivity.

The value of **molar absorptivity is constant** even if the concentration of the solution and the thickness of the container change. But it **differs with different wavelengths**.

Absorbance also varies with wavelength. For this reason, only monochromatic light is used in spectrophotometry and colorimetry.

Molar absorption coefficient ε is expressed as

$$\varepsilon = A^{1\%}_{1cm} \times \frac{\text{Molecular weight}}{10}$$

Where, $A^{1\%}_{1cm}$ is the **specific absorbance which is the absorbance of 1% w/v solution, using a path length of 1cm**. It is a constant for each drug at the given wavelength. This specific absorbance can be used for the quantitative estimation of formulations. ε_{max} is the value of ε at λ_{max}.

DEVIATIONS FROM BEER'S LAW

According to Beer's law, the plot of concentration Vs. absorbance should give a straight line passing through the origin. The straight line can also be obtained by using the line of best fit or method of least squares or by joining the maximum number of points so that the positive and negative errors are balanced or minimized. The regression line obtained can also be used for determining the concentration of a solution. When a nonlinear curve (not a straight line) is obtained in a plot of concentration vs. absorbance, the system is said to have deviations from Beer's law.

Beer's law is normally obeyed only in a certain concentration range, and above or below it may exhibit deviation.

The reasons for the deviations may be classified as

1. Real Deviations
2. Instrumental Deviations
3. Chemical Deviations
4. Incomplete reaction

1. **Real Deviations:** There are two types of real deviations from Beer's law
   (i) **Positive deviation (concave upwards):** It occurs when a small change in concentration produces a great change in absorbance.
   (ii) **Negative deviation (concave downwards):** It occurs when a large change in concentration produces a small change in absorbance.
2. **Instrumental deviations:** Factors like stray radiation, improper slit width, fluctuation in a single beam, and the type of light used can influence the deviation.

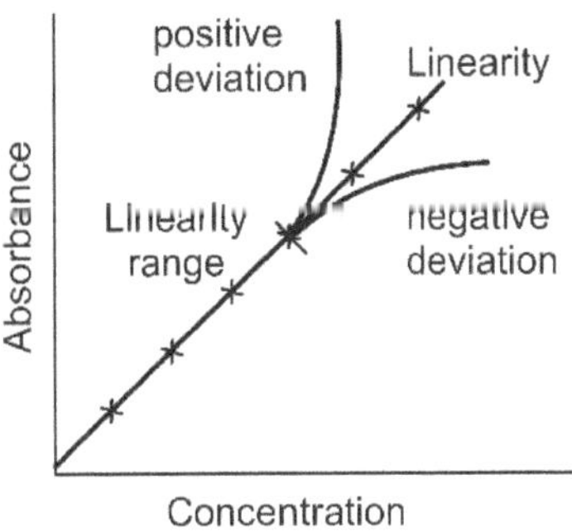

Fig. 1.8 Real deviations

- If monochromatic light is not used, deviation may occur.
- If the width of the slit is not proper and if it allows undesirable radiations to fall on the detector, these undesirable radiations might be absorbed by the impurities present in the sample leading to changes in absorbance.
- The possibility of error due to monochromatic radiation may be minimized by selecting a spectral region, where the change in absorptivity with a change in wavelength is very small.
- This dictates that wavelength selection should be from a broad band (close to λ_{max}) rather than a sharply rising or sharply falling section of the absorption curve.
- It is necessary to prepare a calibrated curve for the absorbance concentration relationship at the chosen wavelength to avoid errors.
3. **Chemical deviations:**
   Physicochemical changes in solution: Factors like association, dissociation, ionization (change in pH), faulty development of colour (incompletion of reaction), and refractive index at high concentration can influence such deviation.
   - **Association:** Methylene blue at a concentration of 10^{-5} M exists as monomer and has λ_{max} of 660nm. But methylene blue at a concentration above 10^{-4} M exist as a dimer or trimer and has a λ_{max} of 600. If the shift of this λ_{max} is not considered during absorbance measurements, then deviations occur.
   - **Dissociation:** Benzyl alcohol in chloroform exists in a polymeric equilibrium;
   $4C_6H_5CH_2OH \leftrightarrow (C_6H_5CH_2OH)_4$.
   Dissociation of the polymer increases with dilution. The monomer absorbs at 2.750 to 2.765 μ, where as the polymer absorbs at 3.0 μ. Hence absorption at 2.75 micron shows a negative deviation whereas at 3.0 μ gives a positive deviation.
   Potassium dichromate at high concentration exists as an orange solution (λ_{max} of 450nm). But on dilution, dichromate ions are dissociated into chromate ions which is yellow in color (λ_{max} of 410nm)
   $$Cr_2O_7^{2-} \text{ (orange)} + H_2O \rightarrow 2H^+ + 2CrO_4^{2-} \text{ (yellow)}.$$

If 450 nm is used for absorbance measurement for dilute solutions, then it leads to deviations from Beer's law.

- **Coloured solute ionization or dissociation** in solution leads to deviation in Beer's law. Colour change of dichromate ion on dilution can be given as an example.
- The deviation may also occur due to the **presence of impurities** that fluoresce or absorb at the absorption wavelength. This interference introduces an error in the measurement of absorption of the sample.
- If the solute undergoes **polymerization**, Ex Benzyl alcohol in carbon tetrachloride in high concentration exists in polymeric form. Dissociation of this polymer increases with dilution, which causes changes in absorbance, causing deviation.
- Beer's law can't be applied for suspensions, but colorimetry can be used with different known concentrations to prepare the reference curve.

4. **Incomplete reaction:**
   - Sufficient time should be allowed before taking absorbance for the reaction to complete so that the colour will be formed completely.
   - If the readings are taken after the colour fades away (instability), deviations can occur. Ex: Limit test for iron: reaction of iron with thioglycolic acid.

INSTRUMENTATION

Table 1.2 Summary of instrumentation

Source of light	Filters and Monochromators	Sample cells	Detectors
Tungsten lamp Carbon arc lamp	Absorption filters. Interference filters Prisms Grating	Glass cuvettes Polystyrene cells for aqueous solutions	1. Barrier layer cell or Photo voltaic cell 2. Photo tubes or Photo emissive cells 3. Photomultiplier tubes.

The common instruments used for measuring emission or absorption of radiant energy are

(i) Photometer (ii) Spectrophotometer

Photometer: These are inexpensive and less accurate instruments. Filters are employed to isolate a wave length region, and a photocell or photo tube is used as a detector. A commercial photometer with filters is called **colorimeter**. It measures either absorbance or transmittance. The working range of wavelength for these instruments is 400-700nm.

Spectrophotometer: These are a little more expensive than colorimeters. They are used for wider wavelength regions, i.e., 360 nm- 900nm or 1000nm. Since these instruments employ **grating monochromators** for scanning and **photomultipliers** as detectors, they are highly accurate. The modern spectrophotometers are microprocessors or computer-based for easy handling of data.

The basic components of both types of instruments are

(a) Source of light

(b) Filters or monochromators

(c) Sample cells

(d) Detectors

Now we will discuss these components

(a) **Source of light /radiation sources: The visible region extends from 400-800nm**

The requirements of the radiation sources are

1. It must be stable and no fluctuations should be there.

2. It should provide continuous radiation from 400-800nm.

3. The intensity should be adequate.

The common radiation sources used in colorimetry (visible region) are

1. Tungsten lamp: Most widely used. The lamp consists of a tungsten filament in a vacuum bulb. It provides sufficient intensity.

Advantages

(i) Stable, robust and easy to use.

(ii) The Emission intensity varies with wavelength.

Disadvantages

(i) The intensity at the shorter wavelength is less

(ii) Difficult to maintain a constant intensity

2. Carbon arc lamp: This lamp is used for high intensity. It provides an entire range of the visible spectrum.

(b) **Filters and Monochromators:** The source emits continuous spectra from 400-800nm. This is called Polychromatic light consisting of several wavelengths. A colorimeter or spectrophotometer works only with a monochromatic light consisting of a single wavelength. A filter or a monochromator is used in these instruments to convert the polychromatic light into monochromatic light.

(a) **Filters:** There are two types of Filters

1. Absorption filters.

2. Interference filters

1. Absorption filters: These filters are made of a solid sheet of glass coloured by dissolved or dispersed pigments. Dyed gelatin is also used as an absorption filter.

The selection of filters is based on this color wheel. If the solution is red coloured, then the green filter is used. For the green solution red filter is used (complementary colors).

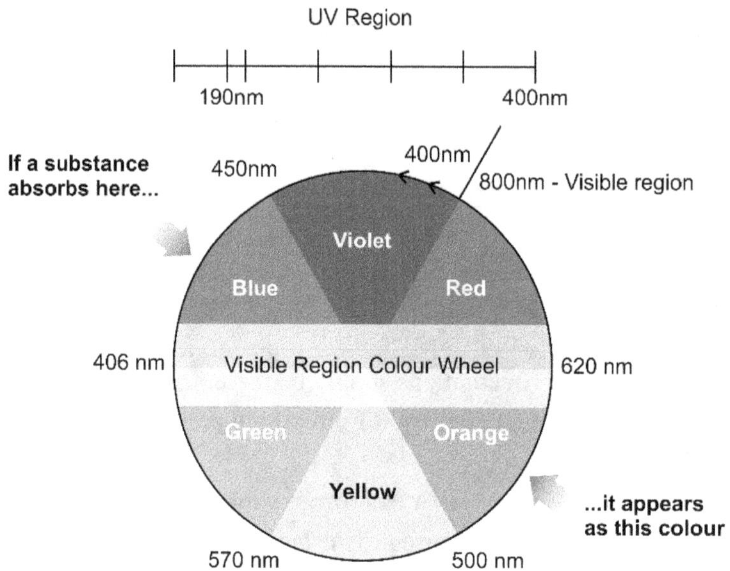

Fig. 1.9 Color wheel showing complimentary colors.

Absorption filters are further classified as **Cut off filters and band pass filters.**

Advantages:

1. Simple construction
2. Cheaper.
3. Easy Filter selection.

Disadvantages:

1. Accuracy is less because of broad band pass ($\pm$ 30 nm).
2. Absorption by filters lead to less intensity of radiation

2. Interference Filters

1. This filter consists of a dielectric spacer film made up of CaF_2, MgF_2, or SiO, between two reflecting parallel silver films.
2. The thickness of the dielectric film may vary to give $1/2\lambda$ (1st order), $2\lambda/2$ (2nd order), $3\lambda/2$ (3rd order spectra), etc.
3. The principle of working is that the radiation reflected by the 2nd film and the incident radiation undergo constructive interference giving monochromatic radiation. The wavelength of the monochromatic radiation depends on the formula

$$\lambda = 2\eta b/m$$

Where λ = wavelength produced

η = dielectric constant of the film material

b = layer thickness

m = order no (1st, 2nd, 3rd, etc)

4. Band pass is 10-15 nm. Transmission is maximum 40%

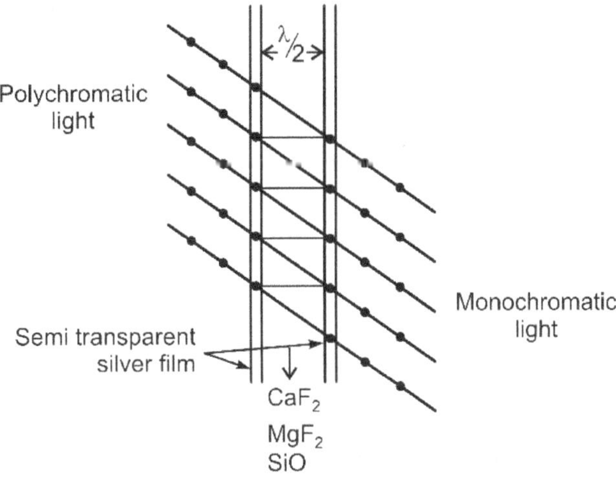

Fig. 1.10 Interference filter – FABRY PEROT.

Advantages:
1. Inexpensive
2. More accurate than absorption filters because of lower band pass.
3. Additional filters can be used to cut off undesired wavelengths.

Disadvantages:
1. Transmission is low
2. Because of the narrow band pass, higher resolution is not obtained.

(b) **Monochromators:** The **monochromators** are more efficient in converting the **polychromatic light to monochromatic light** when compared to filters. The parts of a monochromator are
1. Entrance slit (to get a narrow source of light)
2. Collimator (to make the light parallel)
3. Grating or prism (to disperse the light)
4. Collimator (to reform the monochromatic image of entrance slit)
5. Exit slit (to allow the light to fall on sample cell)

There are two types of monochromators
1. Prisms
2. Grating

1. **Prisms:** The prisms disperse the light into individual wavelengths or colors. They are made of glass and used in inexpensive instruments. Its **resolution is better than filters** because the band width is lower. The resolution of the prism is determined by the size and refractive index of the prism.

There are two types of the prism

I. **Refractive type:** In this type, the light from the source falls on the collimator through the entrance slit. The parallel rays of light from the collimator enter the prism, which disperses the light into component colors or wavelengths. The second collimator receives the dispersed light and reforms the images of the entrance slit. The reformed images are the colors of the VIBGYOR (Violet, Indigo, Blue, Green, Yellow, Orange, and Red). The required wavelength can be selected by rotating the prism or by moving the exit slit and the selected wavelength passes through the exit slit to the sample.

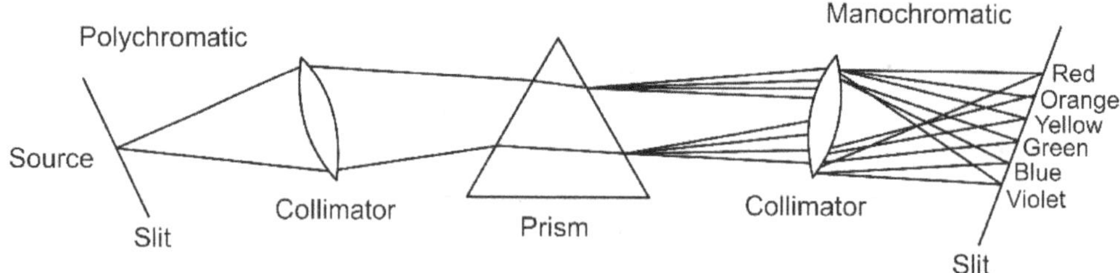

Fig. 1.11 Prism monochromator-dispersive type

II. Reflective type (Littrow mounting): This type has a reflective surface present on one side of the prism. The dispersed radiation gets reflected on the same side of the source and is collected on the same side. The working principle is the same as the refractive type.

The advantage of prisms:

1. Dispersed wavelengths do not overlap

Disadvantages:

1. Non-linear dispersion
2. Temperature sensitive

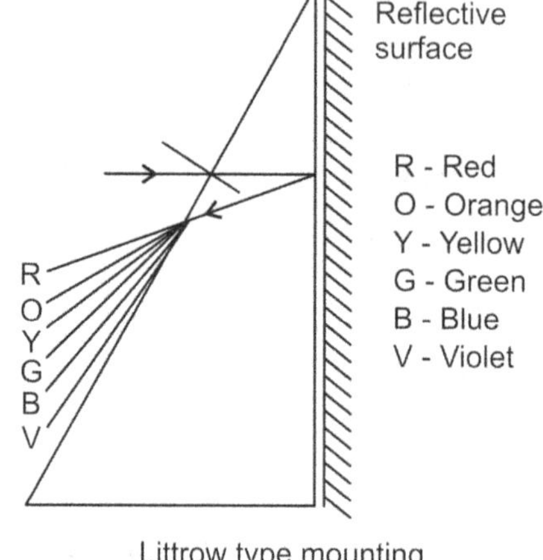

Fig. 1.12 Prism monochromator-reflective type (LITTROW MOUNTING)

2. Gratings: Gratings are **more efficient** in the conversion of polychromatic light to monochromatic light—a **resolution of ±0.1 nm** can be achieved with gratings, and hence they are used in spectrophotometers.

There are two types of Gratings

1. Diffraction grating
2. Transmission grating

1. Diffraction grating: A grating is made up of a large number of parallel lines (grooves) ruled on highly polished surfaces like glass, quartz, alumina, and alkyl halides based on the type of the instrument (visible/UV/IR spectrophotometer). Generally 3600 grooves or more per mm are drawn for ultraviolet and visible regions, and 20 grooves per mm are drawn for the IR spectrophotometer.

These are replica gratings made from the master grating. Replica gratings are prepared by coating the original master grating with epoxy resin and removed after setting. Aluminizing its surface makes the surface of the replica grating reflective.

The principle is **diffraction causes interference**. The incident rays on the grating, get constructive interference with the reflected rays and the resulting radiation wavelength is given by the equation

$$m\lambda = b \,(\sin i \pm \sin r)$$

Where λ = wavelength of produced light

b = grating spacing

i = angle of incidence

r = angle of reflection

m = order (0, 1, 2, 3, etc).

There are two different types of diffraction grating – the ruled grating and the holographic grating.

Gratings are preferred over prisms because of their efficiency.

2. **Transmission grating:** In these types of gratings, refraction takes place. The principle is **refraction produce interferences**. The radiation transmitted through the grating reinforces the partially refracted radiation leading to constructive interferences.

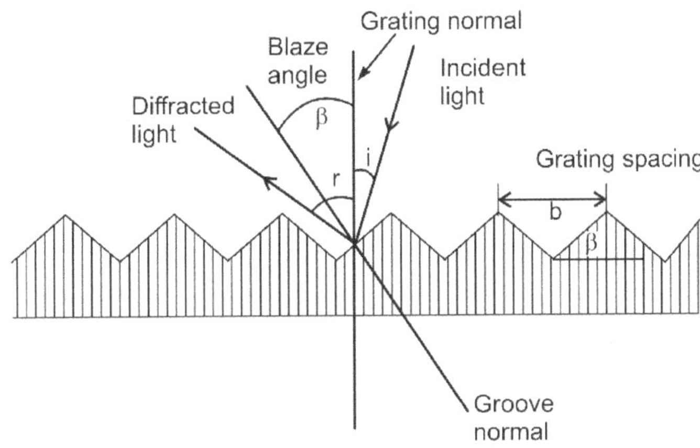

Fig. 1.13 Diffraction grating.

Transmission Grating

The wavelength produced can be calculated using the formula

$$\lambda = \frac{d \sin\theta}{m}$$

Where λ = wavelength of radiation produced.

d = 1/lines per cm

m = order no (0, 1, 2, 3, etc)

θ = angle of deflection/diffraction

Hence, the required light radiation (λ) can be produced either by moving grating and keeping the slit fixed or vice versa.

Advantage:

1. They give linear dispersion.
2. High resolution

Disadvantage:

1. Overlap of spectral orders.

Slits: The two slits are the **entrance slit and the exit slit**. The width of

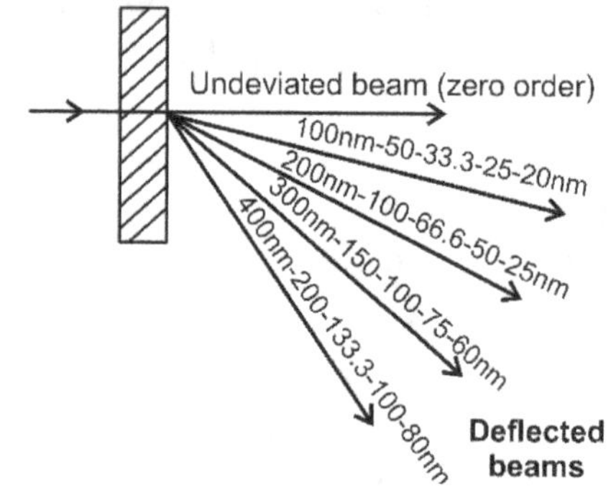

Fig. 1.14 Transmission Grating

the monochromatic image of the entrance slit is the same as the width of the exit slit to pass through the exit slit.

(c) **Sample cells:** The sample cells, known as cuvettes, are used to hold the sample.

The construction material of the cuvette varies with the instrument and the sample nature.

The material of the cuvette should not absorb in the working wavelength.

Material for **visible region: Color corrected fused glass.**

For aqueous solutions, Polystyrene cells are available.

For UV region: Quartz cuvettes are used because **glass absorbs UV radiation**.

The cells are available in different shapes like cylindrical or rectangle.

Small volume cells (0.5ml or less) and large volume cells (5-10ml) are available depending on their capacity to hold sample volume

The common path length (internal distance) is 1cm. Long path length cells with 10cm and short path length cells with 1 or 2mm are also available.

(d) **Detectors:** The radiation from the source through the monochromator passes through the sample. The sample absorbs part of the radiation, and the rest is being transmitted. This transmitted radiation falls on the detector, and the intensity of the absorbed radiation is determined by the detector and displayed.

Detectors used in UV Visible spectrophotometers are called photometric detectors. They convert the **light energy to an electrical signal** which is recorded.

Three types of detectors are

1. Barrier layer cell or Photo voltaic cell
2. Photo tubes or Photo emissive cells
3. Photomultiplier tubes.

1. **Barrier layer cell or Photo voltaic cell:** The detector consists of a metal base like iron or aluminum acting as one electrode. A thin layer of semiconductor material like selenium is deposited on its surface. The surface of the selenium is covered by a very thin layer of silver or gold which acts as a second collector electrode.

The selenium layer separates the two electrodes, and it has extremely low electrical conductivity. When radiation falls on the selenium layer, the electrons are generated at the selenium silver interface. The silver layer collects these electrons. The collection of electrons on the silver surface creates a potential difference between the two electrodes. If the resistance in the external circuit is small, the current flows through the system. The **current produced is directly proportional to the intensity of the incident radiation** on the detector.

Advantages:
 (i)   Simple in design and rugged
 (ii) No external power supply is required.

Disadvantages:
 (i)   Less sensitive to blue region; use is limited to the visible region
 (ii) Amplification of the signal is not possible because the resistance of the external circuit should be low.
 (iii) Fatigue effects.

2. **Photo tubes or Photo emissive cells:** This detector consists of an evacuated glass tube containing a photo cathode and an anode as a collector electrode. The photo cathode is coated with a light-sensitive layer like Cesium, potassium or silver oxide.

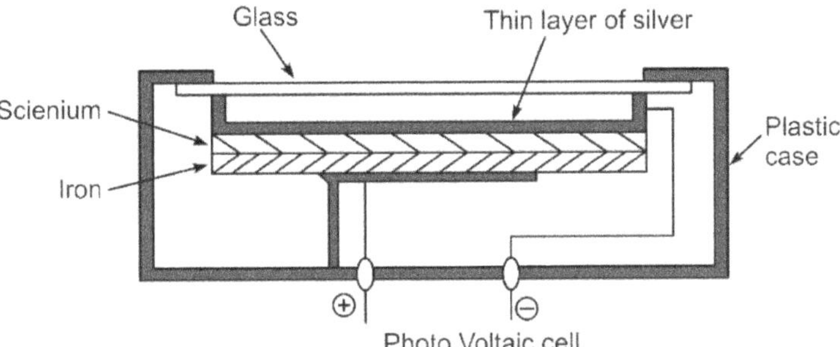

Fig. 1.15 Barrier layer cell.

When the transmitted radiation falls on the photo emissive cathode, it emits photoelectrons. The electron flow towards the anode causes the **current flow, which is proportional to the intensity of radiation falling on the detector.**

Advantages:
1. A signal from the detector can be amplified
2. More sensitive than Photovoltaic cells.
3. Composite coatings containing Cesium, cesium oxide, or silver oxide can be used on the cathode to increase the sensitivity and workable wavelength range.

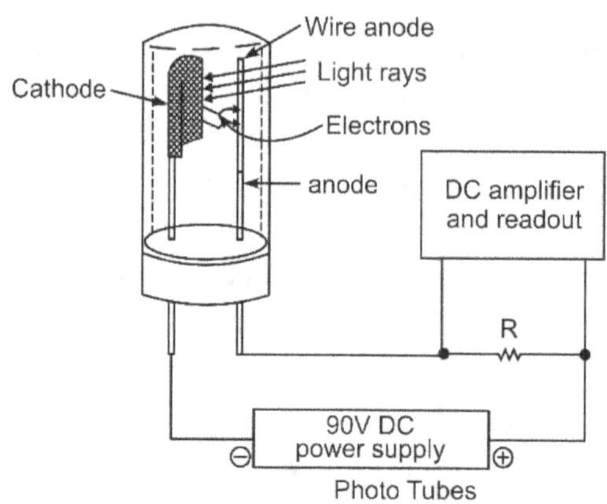

Fig. 1.16 Photo tubes.

3. **Photomultiplier tubes (PMT):** Though expensive, this detector is employed in sophisticated instruments due to its sensitivity.

The working principle of this detector is the **multiplication of photo electrons by secondary emission of electrons.** PMT detector consists of a photo cathode and a series of anodes called dynodes. Each dynode is maintained at a successively higher potential 75-100V higher than the preceding dynode. Up to 10 dynodes are used in the series. When the radiation hits the cathode surface, electrons are emitted. When these electrons hit the dynodes, at each stage, the electrons are multiplied by a factor of 4 or 5 because of the secondary emission. An overall amplification factor of about 10^8 is achieved.

Advantages:

1. Most sensitive

2. Can detect very weak signals

3. Faster response time

4. It can be used in fluorescence measurements

Disadvantages:

1. Expensive

2. It should be shielded from stray light.

Advantages of Colorimetry

1. Specific to one chemical species.

2. Inexpensive per analysis.

3. Even non-chemistry personnel can perform the experiment.

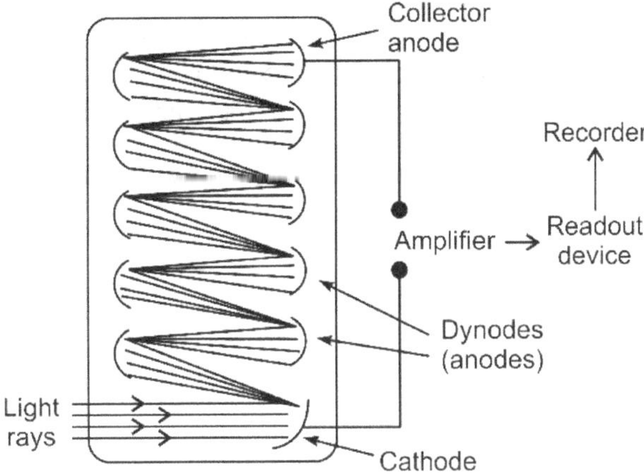

Fig. 1.17 Photo multiplier tube.

Disadvantages

1. Similar colour from interfering substances can produce errors in results
2. May not be suitable for more precise analysis.
3. Matrix interferences produce bad results in controlled conditions.

Types of instruments

The various types of instruments used in colorimetry are Colorimeters, Spectrocolorimeters, and Spectrophotometers. They differ in their construction, cost, sophistication, and automation.

Colorimeters: These instruments contain a tungsten lamp, absorption filters and photovoltaic cells. They read either transmittance or absorbance. Colorimeters are single beam instruments. Wavelength accuracy is normally ± 30nm.

Spectrocolorimeters: Contain prism monochromators instead of filters and either photovoltaic cell or phototube as a detector. These are also single beam instruments measuring either transmittance or absorbance. Wavelength accuracy usually is ± 5nm.

Spectrophotometers: These types of instruments are expensive and more sophisticated. They can read either transmittance or absorbance. They can record the absorption spectrum using a recorder. Spectrophotometers are double beam instruments, where the absorbance of sample and reference solution can be measured simultaneously. The automated devices possess various options like storage of spectrum, comparison of spectra, quantitative techniques, rapid wavelength scanning, data manipulation and derivative spectral mode. Most of them are software-driven instruments. Wavelength accuracy is normally ± 0.1 nm.

These instruments are more accurate and reliable.

The schematic diagram and the working of the instruments are given below

Single beam colorimeter

The source of light here is the tungsten lamp. A concave mirror focuses the light onto the slit. The slit allows the light to pass through the absorption filter, which selects the required wavelength and makes it fall on the sample cell containing the analyte solution. The analyte absorbs the part of the radiation, and the remaining light is transmitted. The intensity of this transmitted radiation is determined by a photo voltaic cell. The diagram of single beam colorimeter is shown in fig 1.18.

Advantages:

1. Simple in construction
2. Less cost
3. Easy operation

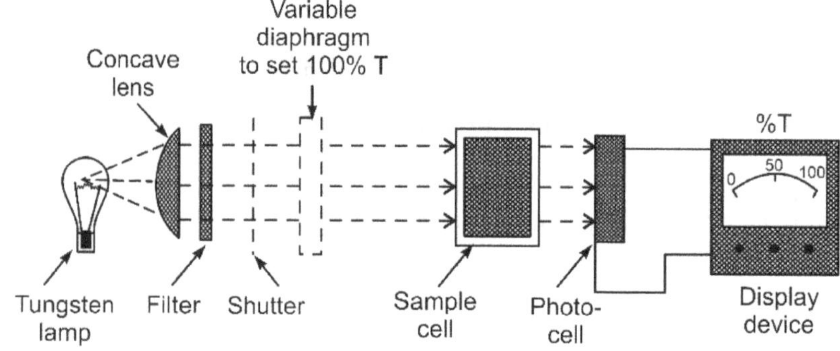

Fig. 1.18 Single beam colorimeter.

Disadvantages:

1. Fluctuations in the intensity of the source affect the absorbance readings.
2. Time-consuming because zero adjustment is to be done in every wavelength.
3. No recorder is available.

Double beam colorimeter: The construction of a double beam colorimeter is the same as a single beam except that the light beam, after passing through the filter or monochromator, is split into sample and reference beams by a beam splitter. Two detectors are used to detect the reference beam and the sample beam individually.

Applications of Colorimetry

1. **Purity determination:** Colorimetry is used to detect the presence of impurities in the sample. Coloured impurities absorb the radiation and give additional peaks in the spectrum.

   E.g., Vitamin B_{12} (cyano cobalamin) absorbs in three wavelengths 278, 361 and 550 nm.

   The ratio of absorbances $\qquad$ 278/361= 0.57

   $\qquad\qquad\qquad\qquad\qquad\qquad$ 550/361 = 0.3

   We can identify the presence of impurities by the change in the absorbance ratios.

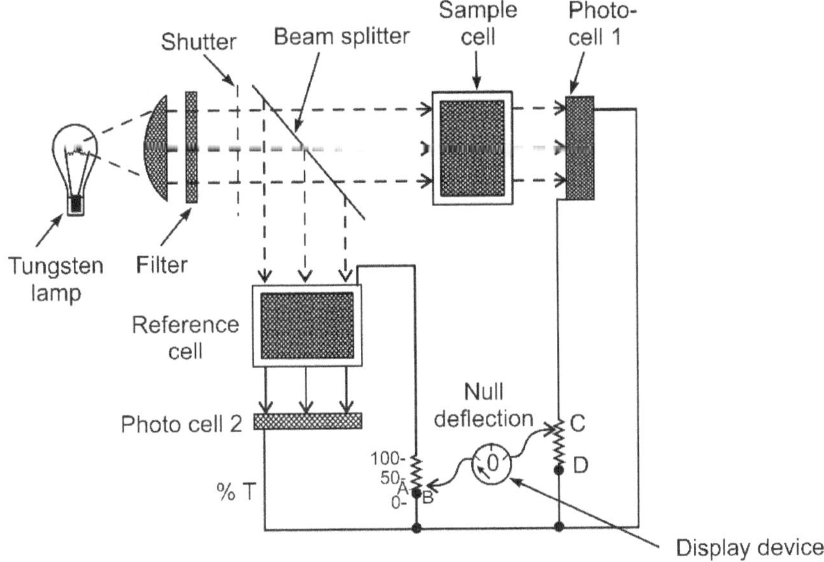

Fig. 1.19 Double beam colorimeter.

2. **Quantitative analysis:** Determination of concentration and amount of the substance present in the given sample is called Quantitative estimation. Using these values, percentage purity can be determined. The quantitative estimation of drugs by colorimetry can be performed by knowing the parameters like λ_{max}, Beer's law concentration range for that substance, solvent, reagents and other conditions.

The types of quantitative estimation are

(a) **Using $A^{1\%}_{1cm}$ value (specific absorbance):** This method is useful in the estimations of raw material as well as formulations when the reference standard is not available. $A^{1\%}_{1cm}$ value can be obtained from Pharmacopoeia, Text books or Journals

(b) **Reference standard available:** If a reference standard is available, specific absorbance can be found experimentally and the above method can be used. Other methods available are

(i) **Direct comparison method (single standard):** The absorbance of a standard solution of known concentration is compared with the absorbance of the sample solution and the concentration of the sample solution is calculated. From the concentration, the amount and the % purity can be calculated.

(ii) **Calibration curve method or multiple standard method:** The calibration graph method is used to minimize the error introduced in the preparation of solutions or absorbance measurements.

In this method, a calibration curve is plotted with concentration vs. absorbance measured with five or more concentrations of the reference or standard. A straight line passing through the origin is obtained. If a straight line is not obtained, a straight line is obtained through the maximum number of points or by regression analysis leading to a line of best fit. The unknown concentration of the sample can be obtained using the absorbance of the sample and the calibration graph. The extrapolation of the graph gives the unknown concentration and amount, and the % purity of the sample can be calculated.

3. **Determination of Ligand/Metal ratio in metallic complexes:** Some organic compounds form metallic complexes with some metals, and the ratio of the metal complexes can be determined using the calibration graph method.

4. **Structural elucidation of organic compounds:** The structure of the unknown compound can be elucidated by comparing the absorption spectrum with that of the known compound. For e.g.,

These are the structures proposed for Tyrosine. When the absorption spectrum of Tyrosine was obtained, it was found to be similar to the absorption spectrum of Phenol. Hence the structure II is proposed for Tyrosine.

5. **Determination of pK$_a$ of indicators:** Determination of dissociation of acid-base indicators like Methyl orange, Methyl red can be determined by the formula

$$pK_a = pH - \log [\text{ionized}]/[\text{unionized}]$$

The plot of absorbance vs. concentration at different pH is recorded and pK$_a$ is calculated using the equation.

6. **Determination of molecular weights:** Molecular weight is determined using the formula

$$\text{Mol. wt} = \varepsilon/a = \varepsilon \times ct/A$$

where   A = absorbance, c = concentration, t = pathlength and a = absorptivity

7. **Determination of functional groups and elements:** Even low concentrations of ions, elements or functional groups can be determined by using chromogenic (colour producing) reagents.

E.g., Determination of Iron using 1,10 phenanthroline, Ammonia using Nessler's reagent and amino acids using Ninhydrin.

8. **Determination of organic compounds and Pharmaceutical substances:** The pharmaceuticals are also determined by colorimetry using chromogenic reagents for producing colors.

E.g., Phenols determined using Ferric chloride, Betamethasone using Phenyl hydrazine.

1.2 UV SPECTROSCOPY

INTRODUCTION

The coloured compounds absorb in the visible region, most of the colorless compounds absorb in the UV region. The UV radiation wavelength starts from the blue end of the visible light

(4000Å) and extends up to 2000 Å. The UV region is divided into two regions

1. Near UV region - 2000 Å -4000 Å (200-400nm)
2. Far or vacuum UV region – below 2000 Å (200nm).

THEORY/PRINCIPLE OF UV SPECTRA

UV Spectroscopy is the study of the absorption of UV radiation by atoms or molecules. Both visible and UV radiations excite only the **valence electrons.** The molecules absorb UV radiation, and they undergo the transition from the ground state to the excited state. Thus, **electronic excitation** is the principle involved in UV spectroscopy. Because of this, UV spectroscopy is also called **Electronic spectroscopy**.

UV absorption is characteristic for each substance, and it depends on the type of electrons present in the molecule. The absorption intensity follows Beer–Lamberts law, and it is proportional to concentration and path length.

Types of electrons in molecules

Three types of electrons are present in molecules. They are

1. **σ electrons** present in saturated compounds. These electrons show absorption in **vacuum UV** region (<200nm)
2. **π electrons** present in unsaturated compounds. (e.g., double or Triple bonds).
3. **n electrons** are called non bonded electrons and are not taking part in bonding between atoms. e.g., organic compounds containing sulphur, nitrogen, oxygen, or halogens.

All molecules having π electrons, n electrons, or a combination of both electrons absorb in the characteristic wavelength region and undergo the transition from the ground state to the excited state. This characteristic absorption helps in the structural elucidation of molecules by identifying the nature of electrons present.

ELECTRONIC TRANSITIONS

When an atom or molecule **absorbs energy, electrons are promoted** from their ground state to an excited state. When a molecule is **electronically excited,** the electrons move from **bonding to antibonding orbitals.** The π electrons are excited to π* antibonding orbitals, and n electrons are excited to either σ* or π* anti bonding orbitals.

The amount of **energy** required to excite **σ electrons** is very high and hence they do not absorb in the **UV region.** The molecules which possess **π electrons (unsaturated compounds)** and **n electrons (lone pair)** absorb energy in the **UV** and **visible region** and these electrons get excited to the higher energy levels. Thus, the possible transitions are **π-π*, n-π*** transitions.

The possible electronic transitions are shown diagrammatically

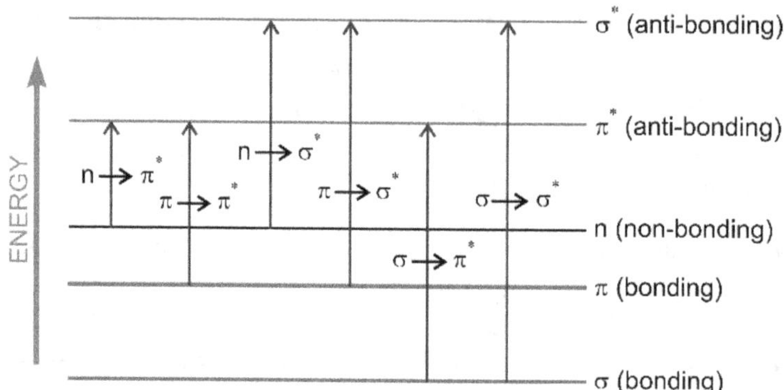

Fig. 1.20 Different types of electronic transitions.

TYPES OF ELECTRONIC TRANSITIONS

The electronic transitions are caused by the absorption of UV light by the molecules leading to the excitation of valence electrons from the ground state to the excited state. The various transitions are **π-π*, n-π*, n-σ* and σ-σ***. The energy requirements for these transitions are given in increasing order n-π*< π-π*< n-σ*< σ-σ*.

Types of transition: It is not necessary that the exposure of the compound to UV radiation will always result in an electronic transition. The probable occurrence of the transition is dependent on the value of the molar extinction coefficient (λ_{max}) and certain other factors. Thus, the transitions can be divided into two types

  (i)   Allowed transitions
  (ii) Forbidden transitions

  (i)   Allowed transitions: Transitions having ε_{max} value 10^4 or more are allowed transitions. They arise due to π-π* transitions.

  E.g., In 1,3 butadiene π-π* transition has ε_{max} value of 21,000 at 217 nm. Thus, it is an allowed transition.

  (ii) Forbidden transitions: Transitions having ε_{max} value less than 10^4 are forbidden transitions. They arise due to n-π* transitions.

  E.g., n-π* transition of saturated aldehyde or ketone shows weak absorption near 290 nm and its ε_{max} value less than 100. So, it is a forbidden transition.

On application of selection rules, σ-σ*, n-σ*,π-π* are allowed transitions, and n-π* transition is forbidden.

DIFFERENT ELECTRONIC TRANSITIONS

  1. n-π*transitions: This transition requires the lowest energy (longer wavelength) of all other transitions. The compounds containing n electrons (present in S, O, N, or halogens) and double or triple bonds (E.g., aldehydes and ketones, nitro compounds) exhibit these transitions. The peaks of these transitions are also called R –bands.

Aldehydes and ketones (RCHO, RCOR) absorption peak,

Without double or triple bonds occurs at 270-300 nm.

With double or triple bonds separated by two or more single bonds occurs at 300-350nm

Identification of n-π*transition can be done by recording the UV spectrum of the substance in an acid medium. The peak due to n-π* transition disappears in acid solution. The presence of other hetero atoms can be identified by comparing the spectrum of a similar compound without hetero atom.

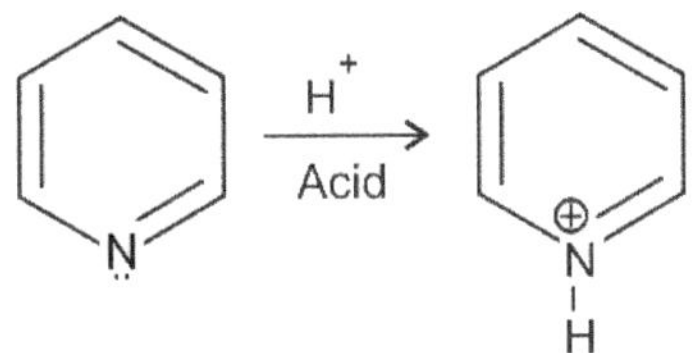

Peak present   Peak disappears

2. **π-π* transitions:** This transition gives B, E and K bands

   B bands are benzenoid bands due to aromatic and hetero aromatic systems. E bands are ethylenic bands due to aromatic systems. K bands (π-π*) are due to conjugated systems. The energy required for this transition lies between n-σ* and n-π*. **The alkyl substitutions and extended conjugations** (addition of alternate double or triple bonds) shift the λ_{max} towards longer wavelength (bathochromic shift). The trans isomer of the alkene absorbs at a longer wavelength with more intensity than the cis isomer (bathochromic shift with hyperchromic effect). Extended conjugation and alkyl substitution shift λ_{max} in such a way that the λ_{max} moves to the colorimetric region. E.g., plant pigments like β carotene, lycopene, etc.

   The λ_{max} of ethylenic (C=C) chromophore is 174nm

   The λ_{max} of acetylenic (C≡C) chromophore is 178 nm

3. **n-σ* transition:**

   Saturated compounds with nonbonding electrons (presence of heteroatoms) exhibit this transition. The energy requirement is less than σ-σ* transition. The wavelength range for this transition is 180-250nm. Transition needs more energy that is shorter wavelength because the lone pair of electrons on the hetero atoms form hydrogen bonds. **Hydrogen bonding shifts UV absorption to a lower wavelength** E.g., of compounds with n-σ* transition Methylene chloride (173nm), water (191nm), Methanol (203nm), Ethanol (204 nm), ether (215 nm) etc.

4. **σ-σ* transitions:**

   This transition requires the highest energy. This transition occurs in saturated compounds (more importantly, hydrocarbons). The absorption peaks appear in the vacuum UV region. 125-135nm. E.g., Methane (122nm), Ethane (135 nm). The commercial UV spectrophotometers operate above 200nm, saturated hydrocarbons like cyclohexane (195 nm) can be used as nonpolar solvents because the solvent peak does not appear.

Laws of absorption

The two laws related to UV Visible spectroscopy are

1. Beer's law (concentration-dependent)
2. Lambert's law (Thickness/path length dependent absorbance)

Lambert's law states that when a beam of monochromatic light is allowed to pass through a transparent medium, the rate of **decrease of intensity** with the **thickness** of the medium is **directly proportional** to the intensity of incident light.

Beer's law: Beer observed a similar relationship between transmittance and the concentration of a solution. Beer's law states that the intensity of a beam of monochromatic light decreases exponentially with an increase in the concentration of absorbing species arithmetically.

$$A = act$$

$$\textbf{Log } I_0/I_t = act$$

is the mathematical expression of Beer Lambert's law,

Where A = absorbance, a = absorption coefficient or absorptivity, c = concentration of the substance and t = path length in cm.

The derivation of Beer lambert's law is given in colorimetry (chapter-1 pg. no. 9-11).

CHROMOPHORES

Any group which exhibits absorption of electromagnetic radiation in the visible or UV region is called as chromophore. It may or may not impart colour to the compound. Example of important chromophores is ethylene, acetylene, carbonyls, acids, esters, and nitrile group.

Eg: NO_2, N=N, C=O, C=C, N=O, C=N, C=C, C=S

Types of Chromophores

1. **Chromophores with π electrons**

    These chromophores undergo $\pi \rightarrow \pi^*$ transitions. E.g., ethylenes and acetylenes.

2. **Chromophores with both π electrons and n (non-bonding) electrons**

    These chromophores undergo two types of transitions.

    $\pi \rightarrow \pi^*$ **and** $n \rightarrow \pi^*$, E.g., carbonyls, nitriles, azo compounds and nitro compounds

Examples of some chromophores

1. Nonconjugated alkenes show intense absorption below 200 nm and are inaccessible to UV spectrophotometer. ($\pi \rightarrow \pi^*$)

2. Non-conjugated carbonyl group compounds give a weak absorption band in the 200-300nm region. ($n \rightarrow \pi^*$).

3. Conjugation of C=C and C=O groups shifts the λ_{max} of both groups to longer wavelengths.

Some simple chromophore groups are given in the Table 1.3.

Table 1.3 Some simple examples of chromophores

Chromophore	Transition	λ_{max}
C=C	$\pi \to \pi^*$	175
C≡C	$\pi \to \pi^*$	175
C=O	$n \to \pi^*$.	285
	$\pi \to \pi^*$	180
N=O	$n \to \pi^*$.	275
	$\pi \to \pi^*$	200
C-X	$n \to \sigma^*$	205
X=BR,I	$n \to \sigma^*$	255

AUXOCHROME

It is a group that itself **does not act as a chromophore. Still,** when **attached to** a chromophore, it shifts the **absorption maximum towards a longer wavelength with an increase in the intensity of absorption.**

Ex: The auxochrome NH_2, when attached to a benzene ring, shifts its absorption maximum (λ_{max}) from 255 to 280nm.

Other Eg: OH, NH_2, OR, NHR and NR_2

All auxochromes have one or more nonbonding pair of electrons.

The auxochrome, when attached to a chromophore, extends the conjugation of chromophores by sharing nonbonding pair of electrons. Ex: $CH_2=CH-:NR_2 \leftrightarrow :^-CH_2 - CH=NR_2^+$

SPECTRAL SHIFTS

Isolated chromophores such as >C=C < and C≡C absorb in the far UV region, which cannot be observed in normal spectrophotometers. The position of absorption maximum and the intensity of absorption can be modified by some structural changes or changes of solvent. These kind of changes of λ_{max} and intensity of absorption are called spectral shifts. The different kinds of spectral shifts are explained below.

BATHOCHROMIC SHIFT OR RED SHIFT

It is the shift of absorption maximum (λ_{max}) **towards longer wavelength** because of the presence of groups like OH and NH_2 (auxochrome) or by change of solvent.

Decreasing the polarity of the solvent causes a redshift in the carbonyl compounds (n-π^* transitions)

When two or more chromophores are present in conjugation in a molecule, redshift is observed.

Eg: ethylene π-π^* transitions   – 170nm

1, 3 butadiene (two double bonds in conjugation)   – 217nm

HYPSOCHROMIC SHIFT OR BLUE SHIFT

It is the shift of absorption maximum towards **shorter wavelength**. This may be caused due to the removal of conjugation in a system or by the change of the solvent. The absorption shift towards a shorter wavelength is also called a **blue shift**.

E.g., Aniline absorption is maximum at 280 nm because the pair of electrons on the nitrogen atom is in conjugation with π bond system of the benzene ring.

In acidic solutions, the absorption shifts to a shorter wavelength (blue shift 200nm). Anilinium ion is formed in acidic solution, and the electron pair is no longer present, and hence conjugation is removed.

HYPER CHROMIC EFFECT

It is the **increase in the intensity of absorption** and is brought about by the introduction of an auxochrome.

E.g.: Introduction of methyl group in 2nd position of Pyridine increases ε_{max} (λ_{max} 262nm) from 2750 to 3560 (λ_{max} 262nm) for π-π^*.

HYPO CHROMIC EFFECT

It is the **decrease in the intensity of absorption** and is brought about by groups that are able to distort the geometry of the molecule.

E.g., When a methyl group is introduced in position 2 of the biphenyl group, there is a decrease in intensity because of distortion caused by the methyl group.

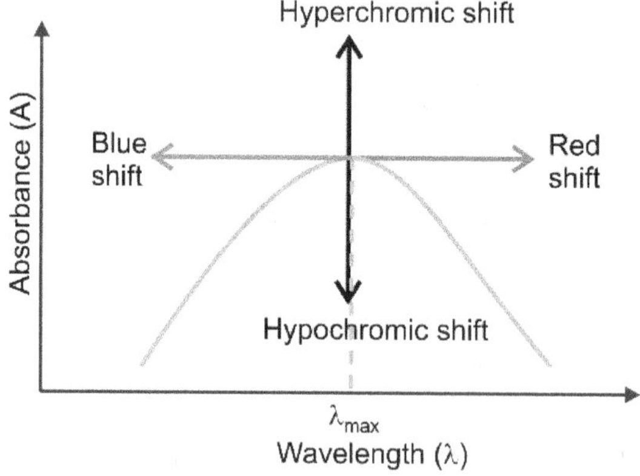

Fig. 1.21 Different types of spectral shifts.

SOLVENT EFFECT ON ABSORPTION SPECTRA

The most commonly used solvent is 95% ethanol. Other solvents which can be used in UV spectroscopy are n-hexane, methyl alcohol, cyclohexane, acetonitrile, diethyl ether, water, etc.

The λ_{max} and the intensity of absorption get shifted for a chromophore by changing the polarity of the solvent. In general, the absorption maximum for the nonpolar compounds is usually shifted with the change in polarity of the solvents.

1. **Effect of solvent on n-π* transitions:** The absorption **shifts to a shorter wavelength** (blue shift) when the **polarity of solvent** is **increased**. In n-π* transitions, (e.g., carbonyl), the ground state is more polar than the excited state. When the polarity of the solvent is increased, the nonbonding electrons in the ground state are more stabilized than the excited state by hydrogen bonding. So, more energy is required to cause excitation, and thus absorption shifts to a shorter wavelength. E.g., absorption maximum of acetone is 279 nm in hexane but it is 264 nm in water.

   Thus, an increase in polarity of solvent generally shifts n-π* and n-σ* transitions to a shorter wavelength.

2. **Effect of solvent on π-π*transitions:** The absorption shifts to a **longer wavelength (red shift)** when the **polarity of solvent is increased.** The π*orbitals get more stabilized by hydrogen bonding with polar solvents like ethanol and water. The polarity of π* orbital is more than the π orbital. When the polarity of the solvent is increased, the dipole-dipole interactions with the solvent molecules lower the energy of the excited state than the ground state. Thus, the energy required for π-π* transitions decreases, leading to shifting of absorption maximum to longer wavelength (red shift).

   Thus, an increase in polarity of solvent generally shifts π-π*transitions to longer wavelength.

   Generally,

1. The absorption maximum will be shifted to a shorter wavelength for a group which is more polar (carbonyl) in the ground state than the excited state (n-π*)
2. The absorption maximum will be shifted to a longer wavelength for a group which is more polar in the excited state than the ground state (π-π*).

CHOICE OF SOLVENT

The solvent for the UV spectroscopy should have the following requirements

   (i) It should not absorb in the region of absorbance measurement of the sample (should be transparent). It should not affect the absorption of the sample.
   (ii) It should be less polar so that its interaction with the solute molecules is minimum.

A commonly used solvent is 95% ethanol. It is cheap, has good solvating power, and is transparent above 210nm.

Example of other solvents transparent above 210 nm is n-hexane, cyclohexane, methanol, water, and ether. The other solvents like benzene, chloroform, and carbon tetrachloride cannot be used because they absorb in the 240-280 nm range.

The solvents like hexane and other hydrocarbons are preferred to polar solvents due to their minimum interactions with the solute molecules.

Some common solvents used in UV spectroscopy and their absorption wavelengths are given in the following Table 1.4.

Table 1.4 Common UV solvents and their absorption wavelengths

Solvent	Wavelength
Water	205nm
Methanol	210nm
ethanol	210nm
Ether	210nm
Chloroform	245nm
Carbon tetra chloride	265nm
Cyclohexene	210nm
Dichloroethane	220nm

INSTRUMENTATION

The common instruments used for measuring the emission or absorption of radiant energy in the UV region are UV spectrophotometers.

Table 1.5 Summary of Instrumentation in UV spectrophotometers

Source of light	Filters and Monochromators	Sample cells	Detectors
Hydrogen discharge lamp Deuterium lamp Xenon Discharge lamp Mercury arc lamp	Absorption filters. Interference filters **Prisms** Refractive type Reflective type **Grating** Diffraction grating Transmission grating	Quartz cuvettes are used because **glass absorbs UV radiation.**	1. Barrier layer cell or Photovoltaic cell 2. Photo tubes or Photo emissive cells 3. Photomultiplier tubes. 4. Photo diode array detector

The essential components of instrument are

(a) Source of light
(b) Filters or monochromators
(c) Sample cells
(d) Detectors
(e) Display or readout devices
(f) Computers and recorders

Now we will discuss these components

(a) Source of light /radiation sources: The Ultraviolet (UV) region extends from 200-400nm

The requirements of the radiation sources are

1. It must be stable, and no fluctuations should be there.

2. It should provide continuous radiation from 200-400nm.
3. Its intensity should be adequate.

   (i) Hydrogen discharge lamp: Hydrogen gas is stored under high pressure in these lamps. When an electric discharge is passed through the gas, excited hydrogen molecules are produced, which emit UV radiations. The high pressure of hydrogen gives a continuous spectrum. Hydrogen lamp gives radiation from 120 -350 nm.

   Advantages:
   1. Stable, robust and widely used
   2. Emits broadband (continuous spectrum)

   (ii) Deuterium lamp: In this lamp, deuterium is used in the place of hydrogen, leading to 3 to 5 times increase in the intensity of emitted radiation.

   Advantages:
   1. More intensity

   Disadvantage
   1. Expensive

   (iii) Xenon Discharge lamp: Xenon gas is stored under 10-30 atmosphere pressure in this lamp. It has got two tungsten electrodes separated by a distance of 8 mm. When a low voltage is applied, an intense arc is produced between the electrodes producing UV radiation. The intensity produced is greater than the hydrogen discharge lamp.

   (iv) Mercury arc lamp: In this lamp, mercury vapour is used under high pressure, and the mercury atoms are excited by electrical discharge. The spectrum produced is not continuous and hence not used widely.

(b) Filters and Monochromators: The source emits continuous spectra from 200-400nm. This is called polychromatic light consisting of several wavelengths. A colorimeter or spectrophotometer works only with a monochromatic light consisting of a single wavelength. A filter or a monochromator is used in these instruments to convert the polychromatic light into monochromatic light.

   (i) Filters: There are two types of filters
   1. Absorption filters.
   2. Interference filters

   1. Absorption filters: These filters are made of a solid sheet of glass coloured by dissolved or dispersed pigments. Dyed gelatin is also used as an absorption filter.

   Absorption filters are further classified as Cut off filters and bandpass filters.

   Advantages:
   1. Simple construction
   2. Cheaper.
   3. Easy Filter selection.

Disadvantages:
1. Accuracy is less because of broad bandpass ($\pm$ 30 nm).
2. Absorption by filters lead to less intensity of radiation

2. Interference Filters

1. This filter consists of a dielectric spacer film made up of CaF_2, MgF_2, or SiO, between two reflecting parallel silver films.
2. The thickness of the dielectric film may vary to give $1/2\lambda$ (1st order),$2\lambda/2$ (2nd order),$3\lambda/2$ (3rd order), etc.

   The principle of working is that the radiation reflected by the 2nd film and the incident radiation undergo constructive interference to give monochromatic radiation. The wavelength of the monochromatic radiation depends on the formula

$$\lambda = 2\eta b/m$$

   Where λ = wavelength produced

   η = dielectric constant of the film material

   b = layer thickness

   m = order no(1st, 2nd, 3rd, etc)
3. Bandpass is 10-15 nm. Transmission is maximum 40%

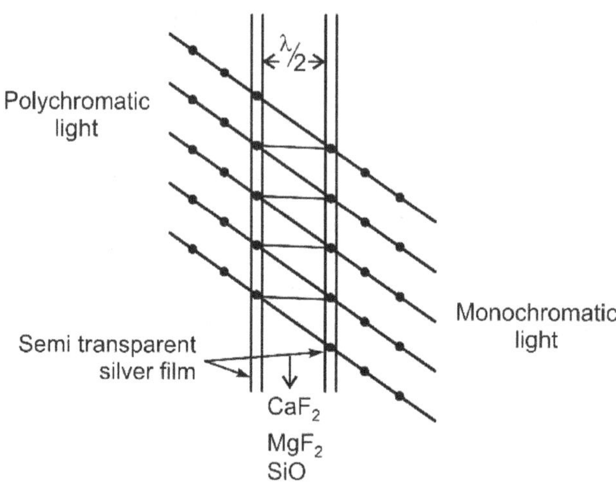

Fig. 1.22 Interference filter FABRY PEROT.

Advantages:
1. Inexpensive
2. More accurate than absorption filters because of the lower bandpass.
3. Additional filters can be used to cut off undesired wavelengths.

Disadvantages:

1. Transmission is low
2. Because of the narrow bandpass, higher resolution is not obtained.

(ii) Monochromators: The **monochromators** are more efficient in converting the **polychromatic light to monochromatic light** when compared to filters. The parts of a monochromator are

1. Entrance slit (to get a narrow source of light)
2. Collimator (to make the light parallel)
3. Grating or prism (to disperse the light)
4. Collimator (to reform the monochromatic image of entrance slit)
5. Exit slit (to allow the light to fall on sample cell)

There are two types of monochromators

1. Prisms
2. Grating

1. **Prisms:** The prisms disperse the light into individual wavelengths or colours. They are made of glass and used in inexpensive instruments. Its resolution is better than filters because the bandwidth is lower. The resolution of the prism is determined by the size and refractive index of the prism.

   There are two types of the prism

   I. **Refractive type:** In this type, the light from the source falls on the collimator through the entrance slit. The parallel rays of light from the collimator enter the prism, which disperses the light into component colours or wavelengths. The second collimator receives the dispersed light and reforms the images of the entrance slit. The reformed images are the colours of the VIBGYOR (Violet, Indigo, Blue, Green, Yellow, Orange, and Red). The required wavelength can be selected by rotating the prism or by moving the exit slit and the selected wavelength passes through the exit slit to the sample.

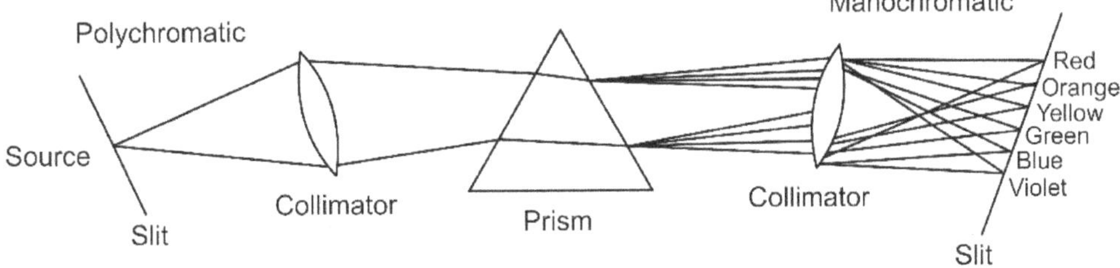

Fig. 1.23 Prism monochromator-Dispersive type.

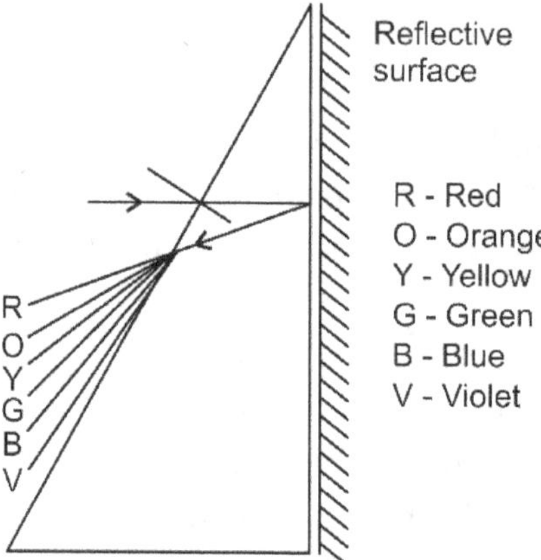

Fig. 1.24 Prism monochromator-Reflective type (Littrow mounting).

II. Reflective type (Littrow mounting): This type has a reflective surface present on one side of the prism. The dispersed radiation gets reflected on the same side of the source and is collected on the same side. The working principle is the same as the refractive type.

The advantage of prisms:

1. Dispersed wavelengths do not overlap

Disadvantages:

1. Non-linear dispersion
2. Temperature sensitive

2. **Gratings:** Gratings are more efficient in the conversion of polychromatic light to monochromatic light. A resolution of ±0.1 nm can be achieved with gratings, and hence they are used in spectrophotometers.

There are two types of Gratings

I. Diffraction grating

II. Transmission grating

I. **Diffraction grating:** A grating is made up of a large number of parallel lines(grooves) ruled on highly polished surfaces like glass, quartz, alumina and alkyl halides based on the type of the instrument (visible/UV/IR spectrophotometer. Generally, 3600 grooves or more per mm are drawn for ultraviolet and visible regions, and 20 grooves per mm are drawn for the IR spectrophotometer.

These are replica gratings made from the master grating. These are prepared by coating the original master grating with epoxy resin and removed after setting. Aluminizing its surface makes the surface of the replica grating reflective.

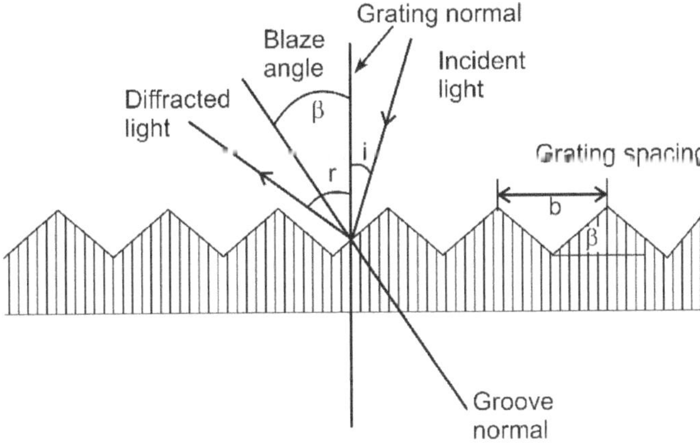

Fig. 1.25 Diffraction grating.

The principle is diffraction causing reinforcement. The incident rays on the grating, get reinforced with the reflected rays and the resulting radiation wavelength is given by the equation

$$m\lambda = b\,(\sin i \pm \sin r)$$

Where λ = wavelength of produced light

           b = grating spacing

           i = angle of incidence

           r = angle of reflection

           m = order (0, 1, 2, 3, etc).

Gratings are preferred over prisms because of their efficiency.

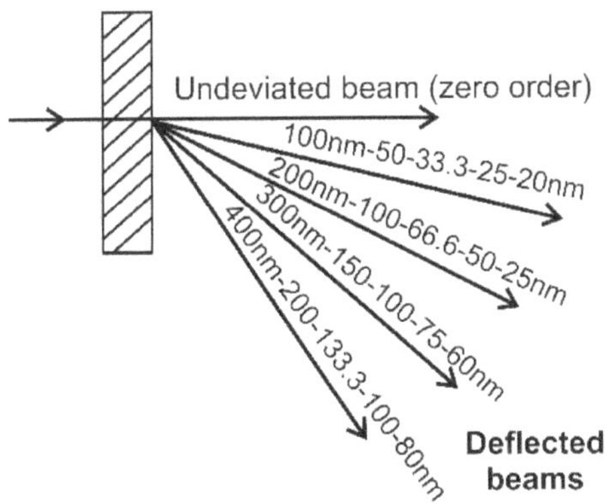

Fig. 1.26 Transmission grating.

II. **Transmission grating:** In these types of gratings, refraction takes place. The principle is refraction produces reinforcement. The radiation transmitted through the grating reinforces with the partially refracted radiation leading to reinforcements.

Transmission grating: The wavelength produced can be calculated using the formula

$$\lambda = \frac{d \, \sin\theta}{m}$$

Where $\quad$ λ = wavelength of radiation produced.

d = 1/lines per cm

m = order no (0, 1, 2, 3, etc)

θ = angle of deflection/diffraction

Hence, the required light radiation (λ) can be produced either by moving grating and keeping the slit fixed or vice versa.

Advantage:

1. They give linear dispersion.
2. High resolution

Disadvantage:

1. Overlap of spectral orders.

Slits: The two slits are the **entrance slit and the exit slit.** The width of the monochromatic image of the entrance slit is the same as the width of the exit slit so that it can pass through the exit slit.

(c) **Sample cells:**

- The sample cells, known as cuvettes, are used to hold the sample.
- The construction material of the cuvette varies with the instrument and the sample nature.
- The material of the cuvette should not absorb in the working wavelength.
- Material for **visible region**:

  Colour corrected fused glass.

  For aqueous solutions, Polystyrene cells are available.

  For UV region: Quartz cuvettes are used because **glass absorbs UV radiation.**
- The cells are available in different shapes like cylindrical or rectangle.
- Small volume cells (0.5ml or less) and large volume cells (5-10ml) are available depending on their capacity to hold sample volume
- The common path length (internal distance) is 1cm. Long path length cells with 10cm and short path length cells with 1 or 2mm are also available.

(d) **Detectors:** The radiation from the source through the monochromator passes through the sample. The sample absorbs part of the radiation, and the rest is being transmitted. This transmitted radiation falls on the detector, and the intensity of the absorbed radiation is determined by the detector and displayed.

Detectors used in UV Visible spectrophotometers are called as photometric detectors. They convert the **light energy to an electrical signal** which is recorded.

Three types of detectors are

1. Barrier layer cell or Photovoltaic cell
2. Photo tubes or Photo emissive cells
3. Photomultiplier tubes.

1. **Barrier layer cell or Photo voltaic cell:** The detector consists of a metal base like iron or aluminium acting as one electrode. A thin layer of semiconductor material like selenium is deposited on its surface. The surface of the selenium is covered by a very thin layer of silver or gold which acts as a second collector electrode.

   The selenium layer separates the two electrodes, and it has extremely low electrical conductivity. When radiation falls on the selenium layer, the electrons are generated at the selenium silver interface. The silver layer collects these electrons. The collection of electrons on the silver surface creates a potential difference between the two electrodes. If the resistance in the external circuit is small, the current flows through the system. The **current produced is directly proportional to the intensity of the incident radiation** on the detector.

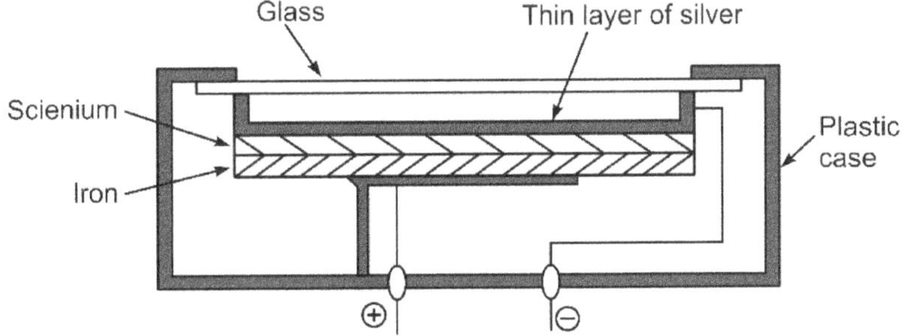

Fig. 1.27 Barrier layer cell.

Advantages:
1. Simple in design and rugged
2. No external power supply is required.

Disadvantages:
1. Less sensitive to blue region. use is limited to the visible region
2. Amplification of the signal is not possible because the resistance of the external circuit should be low.
3. Fatigue effects.

1. **Photo tubes or Photo emissive cells:** This detector consists of an evacuated glass tube containing a photocathode and an anode which acts as a collector electrode. The photocathode is coated with a light-sensitive layer like Cesium, potassium, or silver oxide. When the transmitted radiation falls on the photoemissive cathode, it emits photoelectrons. The electron flow towards the anode causes the **flow of the current, which is proportional to the intensity of radiation falling on the detector.**

Advantages:
1. The signal from the detector can be amplified
2. More sensitive than Photovoltaic cells.
3. Composite coatings containing Cesium, caesium oxide, or silver oxide can be used as the cathode to increase the sensitivity and workable wavelength range.

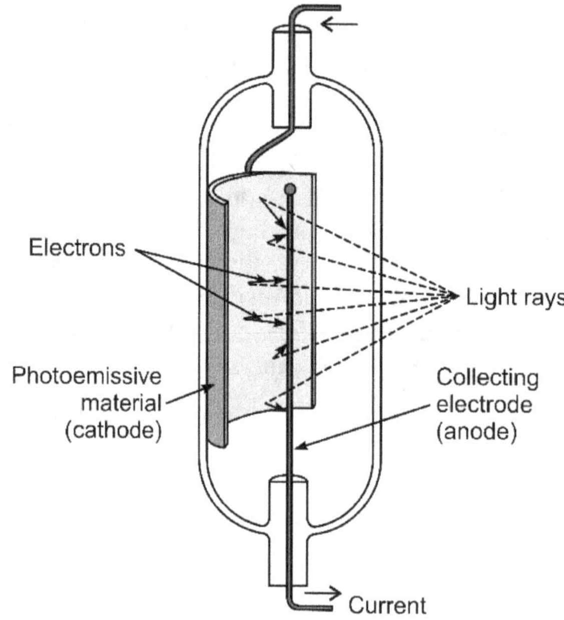

Fig. 1.28 Photo tube.

2. **Photomultiplier tubes (PMT):** Though expensive, this detector is employed in sophisticated instruments due to its sensitivity.

The working principle of this detector is the **multiplication of photoelectrons by secondary emission of electrons.** PMT detector consists of a photocathode and a series of anodes called dynodes. Each dynode is maintained at a successively higher potential 75-100v higher than the preceding dynode. Up to 10 dynodes are used in the series.

When the radiation hits the cathode surface, electrons are emitted. When these electrons hit the dynodes, at each stage, the electrons are multiplied by a factor of 4 or 5 because of the secondary emission. An overall amplification factor of about 10^8 is achieved.

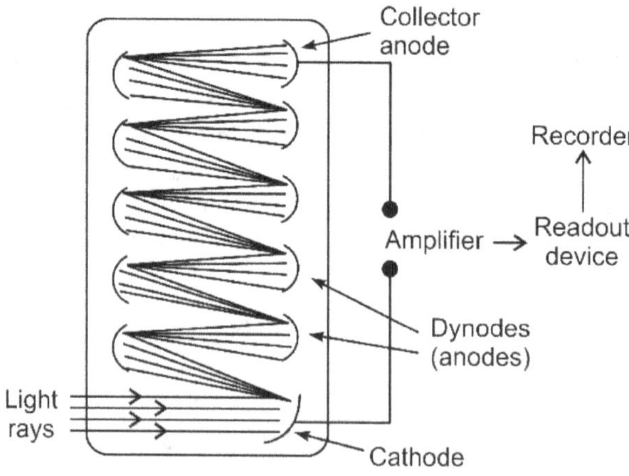

Fig. 1.29 Photo multiplier tube (PMT).

Advantages:

1. Most sensitive
2. Can detect very weak signals
3. Faster response time
4. It can be used in fluorescence measurements

Disadvantages:

1. Expensive
2. It should be shielded from stray light.

4. Silicone photodiode detector: (PDA detector):

Photodiodes are made up of photosensitive semiconductor materials like silicon, gallium, arsenide, etc. They absorb light in the wavelength range 250 to 1100 nm, characteristic of silicon. These detectors can measure light at different wavelengths simultaneously.

A photodiode is one type of light detector used to convert the light into current.

It comprises of optical filters, built-in lenses and also surface areas. These diodes have a slow response time when the surface area of the photodiode increases.

But they are less sensitive than PMTs.

Working of Photodiode

The working principle of a photodiode is, when a photon of ample energy strikes the diode, it makes a couple of electron-hole. This mechanism is also called the inner photoelectric effect. If the absorption arises in the depletion region (junction), then the carriers are removed from the junction by the inbuilt electric field of the depletion region. Therefore, holes (positive charge where the electrons are removed) in the region move toward the anode, electrons move toward the cathode, and a photocurrent will be generated.

A photodiode continually operates in a reverse bias mode. The current produced is directly proportional to light incident on the Photodiode layer

PDA detector/Diode Array Detectors (DAD)

It is an assembly of individual photodiodes on a chip.

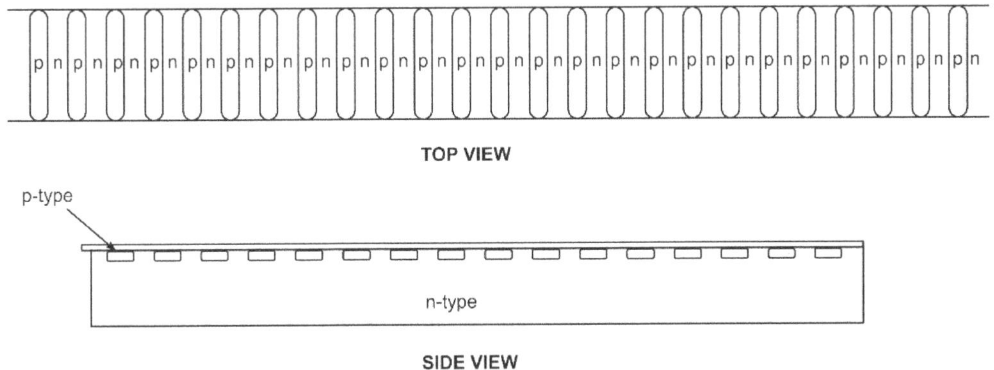

Fig. 1.30(a) Assembly of individual photodiodes on a chip.

Each diode can be addressed individually

Monochromator disperses light across PDA, with a small no of diodes /wavelength.

Allow simultaneous collection of all wavelengths.

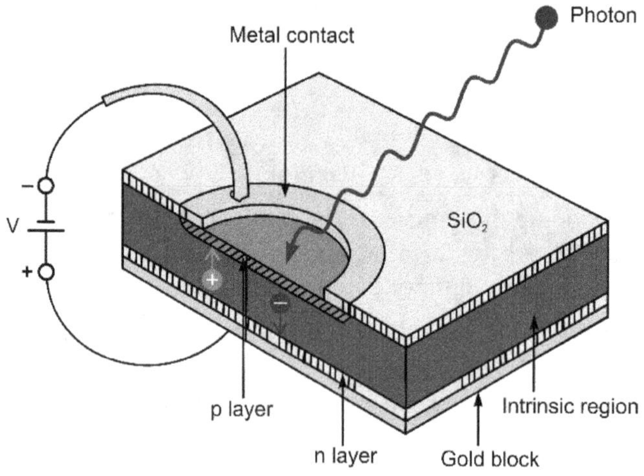

Fig. 1.30(b) Photo diode array (PDA) detector.

DADs differ from UV-VIS detectors in that **light from the lamps is shone directly onto the flow** cell, light that passes through the flow cell is **dispersed by the diffraction grating,** and the amount of the dispersed light is estimated for each wavelength in the photodiode arrays

Advantages of PDA detector

1. Determination of absorbance in all wavelengths in one run.
2. Absorbance maximum can be calculated for each peak
3. Peak purity can be determined
4. Rapid scanning.
5. Used for kinetic studies
6. Used as a detector in LC

Disadvantages

1. Large noise because the amount of light is small
2. Lamp fluctuations

(e) **Display or Readout Devices:** The electrical energy from the detector is displayed in a read-out system such as LCD. They have faster responses and are easier to read.

(f) **Computers and Recorders:** Computers are helpful in storing and processing of the data received from the detector. The recorders are for recording the absorption spectrum or absorbance of the analyte of interest.

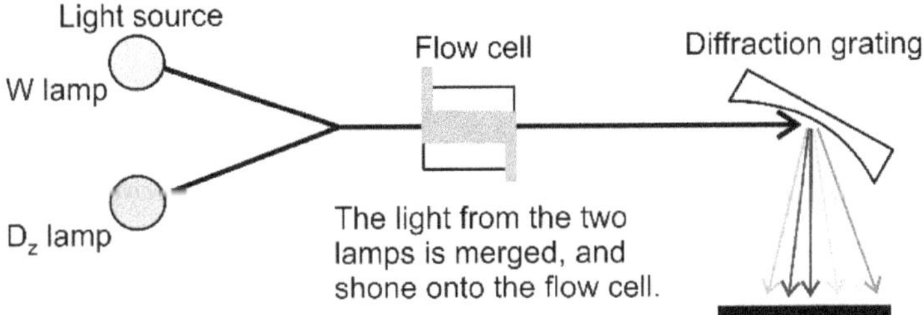

Fig. 1.31 Working of photo diode array (PDA) detector.

The different types of instruments used in UV spectroscopy are
1. Single Beam UV spectrophotometer
2. Double beam UV spectrophotometer

1. Single beam spectrophotometer: The construction and working of a single-beam spectrophotometer are the same as a single beam colorimeter, except the source and the detectors used are different here. The schematic diagram is given below

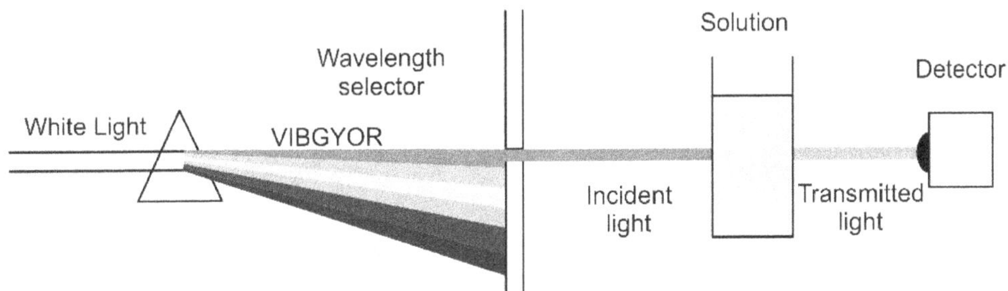

Fig. 1.32 Single beam spectrophotometer.

2. Double beam UV Spectrophotometer: The construction and working of the double beam are the same as that of the double beam colourimeter. The difference is that instead of a filter or prism, here grating monochromator is used, and photomultiplier detectors are used in the place of photo tube detectors. These are the widely used instruments.

Working:
1. The radiation from the source passes via the mirror system to the monochromator.
2. The monochromator allows a narrow range of wavelengths to pass through the exit slit.
3. The radiation from the monochromator through the exit slit is received by the optical chopper, which divides the beam into two beams; one passes through the reference and the other through the sample cell.
4. The beams, after passing through reference and sample cells, are passed to the detector.

5. The output of the detector is connected to the amplifier, which responds to the change in transmission through the sample and reference.
6. Then the signal is transferred to the recorder, where the absorbance or transmittance is recorded as a function of wavelength.

Advantages:

1. Zero adjusting can be done for all wavelengths simultaneously and need not be repeated for every wavelength.
2. In double beam spectrophotometers, the ratio of sample and reference beams is used for measurement, minimizing the errors due to variation in the intensity of the source and detector fluctuation.
3. Rapid scanning over a wide wavelength region.
4. High accuracy, sensitivity, reliability, and repeatability.

Disadvantages:

1. Complicated
2. Expensive

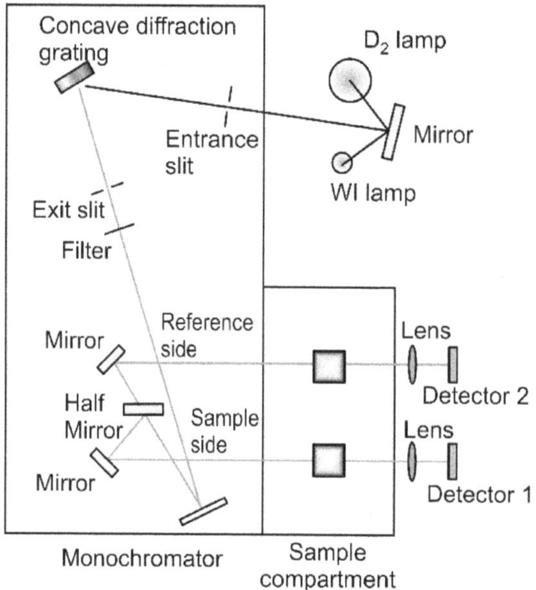

Fig. 1.33 Double beam UV Visible spectrophotometer.

APPLICATIONS

1. **Spectrophotometric titrations:** In these titrations, the absorbance of the solution is measured after each addition of the titrant, and the **absorbance is plotted against the volume of the titrant.** The endpoint is determined from the graph. The titrant or the titrated substance can be the absorbing species. The endpoint is determined by the

intersection of lines of the titration curve. If there is a curvature without intersection due to incomplete reaction, the lines can be extrapolated to give the endpoint

The first curve (a) is the example where the **titrant only absorbs**. In this titration, only the titrant that is bromine is absorbing, and its absorption is plotted against the volume of the titrant. So, the absorbance will not change till all the arsenic is consumed. Once all the arsenic is consumed by the titrant, only bromine will be there, causing an increase in absorption.

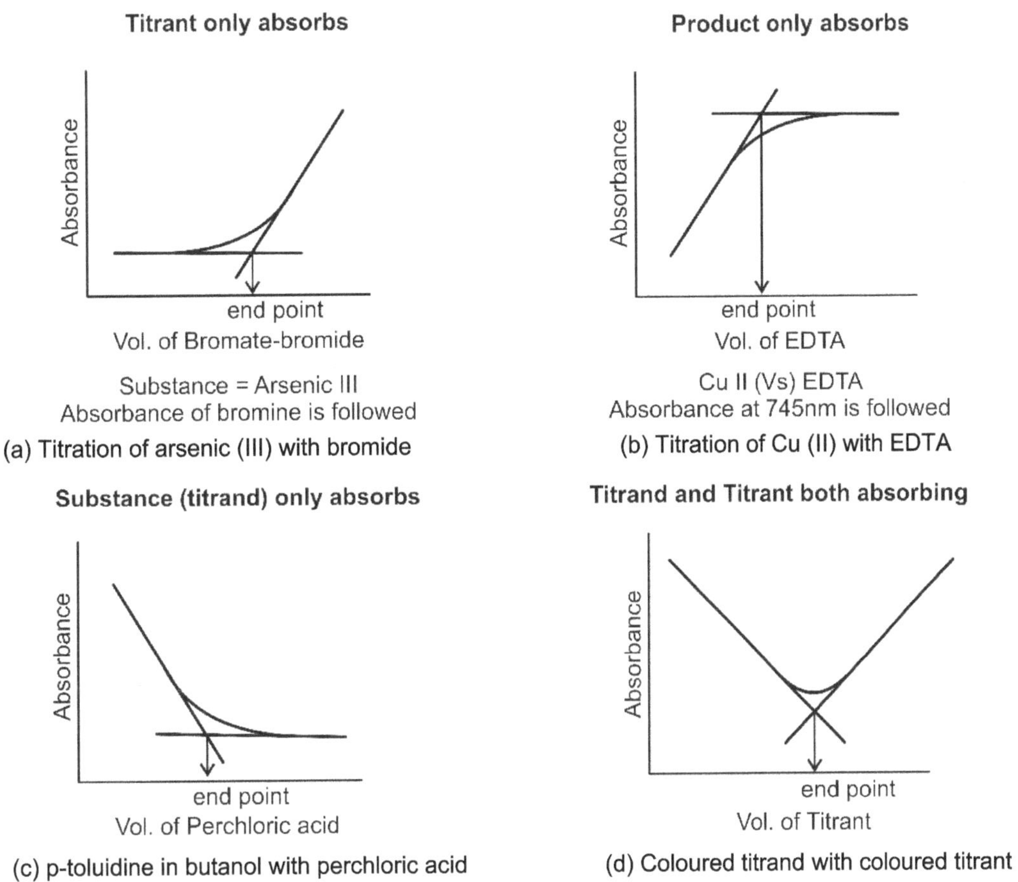

Fig. 1.34 Types of spectrophotometric titrations.

The second curve (b) is the example of the case where **only the product is absorbing**. E.g., is the titration of Cu II ions with Ethylene diamine tetracetic acid (EDTA). There is an increase in absorbance as the product is formed, and once the reaction is complete, no new product is formed, and the absorbance becomes constant with no further increase.

In the third case, (c) **only the substance titrated (titrand) absorbs.** Eg is the titration of p-toluidine in butanol with perchloric acid. As p-toluidine is being consumed by the perchloric acid, there is a decrease in absorbance, and once all the p-toluidine is consumed, there is no change in absorbance. The graph becomes horizontal with no further change in absorbance.

In the fourth case (d), coloured substance reacts with coloured titrant to give a colourless product. So, **both the titrand and the titrant will be absorbing and the product is not absorbing**. Initially, there is a sharp decrease in absorbance due to the formation of non-absorbing product formation. After the endpoint, the concentration of the absorbing titrant increases, leading to an increase in absorbance.

Conditions to be followed:

(a) Titrand, titrant, or product should absorb in the selected wavelength.

(b) The concentration should be in the Beer's law range

(c) Titrant must be stronger so that dilution error can be minimized (absorbance is proportional to concentration).

Applications

1. **Acid-base titrations:** Eg: titration of phenols with sodium hydroxide. Absorbance due to phenolate ion is recorded.

2. **Oxidation-reduction titrations:** Eg: Ce (III) titration with Co (III).Ce(IV) formation is followed.

3. **Complexometric Titrations:** Eg: Cu (II) is titrated with EDTA. Cu-EDTA complex formation is followed.

4. **Precipitation Titrations:** Eg: SO_4^{2-} ions titrated with Ba (II) ions, and the turbidity appearance is followed.

Advantages

1. As only one absorbing species is needed, these titrations can be applied to a large no. of non –absorbing substances.

2. No interference from other absorbing species, as only the change in absorbance is important

3. The method is applicable to highly coloured solutions where visual indicators cannot be used.

4. The endpoint can be determined even if the reaction is incomplete

5. Accurate and precise.

6. No need to work only at the absorption maximum (λ_{max}). So wavelength choice is there.

7. Dissociation constants can be determined from these graphs.

8. Convenient, rapid.

2. **Single component analysis:** Determination of concentration and amount of the substance present in the given sample is called Quantitative estimation. Using these values, percentage purity can be determined. The quantitative estimation of drugs by UV-visible spectrophotometry can be performed by knowing the parameters like λ_{max}, Beer's law concentration range for that substance, solvent, reagents and other conditions.

The types of quantitative estimation are

1. **Using $A^{1\%}_{1cm}$ value (specific absorbance):** This method is useful in the estimations of raw material as well as formulations when the reference standard is not available. $A^{1\%}_{1cm}$ value can be obtained from Pharmacopoeia, Textbooks or Journals

2. **Reference standard available:** If reference standard is available, specific absorbance can be found experimentally and the above method can be used.

3. **Direct comparison method (single standard):** The absorbance of the standard solution of known concentration is compared with the absorbance of sample solution and the concentration of sample solution is calculated. From the concentration, the amount and the % purity can be calculated.

4. **Calibration curve method or multiple standard Method:** The calibration graph method is used to minimize the error introduced in the preparation of solutions or absorbance measurements.

   In this method, a calibration curve is plotted with concentration vs. absorbance measured with five or more concentrations of the reference or standard. A straight line passing through the origin is obtained. If a straight line is not obtained, a straight line is obtained through the maximum no of points or by regression analysis leading to a line of best fit. The unknown concentration of the sample can be obtained using the absorbance of the sample and the calibration graph. The extrapolation of the graph gives the unknown concentration and amount and the % purity of the sample can be calculated.

5. **Multi-component analysis:** Simultaneous analysis of numerous components in a sample is called multi-component analysis. Several methods are available in UV spectroscopy for multi-component analysis. In these methods, the absorption spectra is recorded and mathematically processed.

The basis of all the Spectrophotometric techniques for multi-component samples is "At all wavelengths, the absorbance of a solution is the sum of absorbance of the individual components"

or

The measured absorbance is the difference between the total absorbance of the solution in the sample cell and that of the solution in the reference cell.

The excipients present in the formulation should not be absorbed at the wavelength of the experiment.

The different UV spectrophotometric multi component analysis methods are

1. Simultaneous equation method (Vierdott's method)
2. Absorbance ratio method (Q-Absorbance method)
3. Derivative Spectrophotometric method
4. Multi wavelength UV-spectrophotometry
5. Dual wavelength method
6. Difference spectroscopy
7. Geometric correction method
8. Orthogonal poly nominal method

9. Solvent extraction method
10. H-point standard addition method
11. Least square approximation method

1. **Simultaneous equation method (Vierdott's method):** If a sample contains two absorbing drugs (X & Y) with λ_{max} at λ_1 and λ_2 respectively, then it is possible to determine the concentration of both the drugs by simultaneous equation method if certain criteria are satisfied.

   Criteria for using the simultaneous equation method are

   The criteria are that the ratios,

   $$\frac{A_2 / A_1}{ax_2 / ax_1} \quad \text{and} \quad \frac{ay_2 / ay_1}{A_2 / A_1}$$

   should lie outside the range of 0.1-2

   This criteria is satisfied only when the λ_{max} of the two components are reasonably dissimilar.

   The second criterion is that the two components should not interact chemically with each other.

   Example: Validated spectrophotometric methods for simultaneous' estimation of Telmisartan and Indapamide in the pharmaceutical dosage form. λ_{max}: TEL 296 nm IND 242 nm

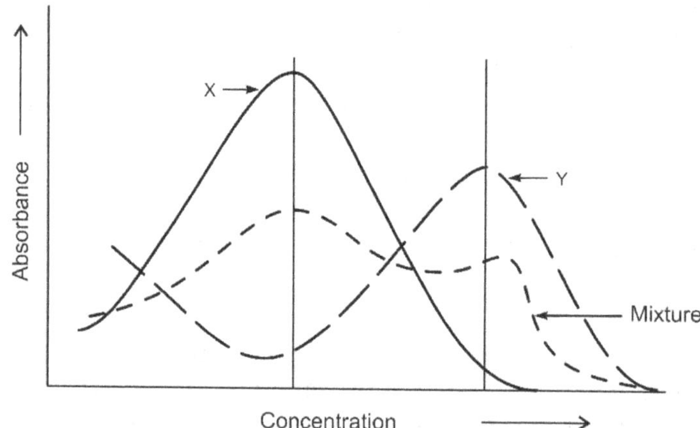

Fig. 1.35 Simultaneous estimation of Telmisartan and Indapamide in pharmaceutical dosage forms.

At λ_1 and λ_2, the absorbance of the mixture is the sum of the individual absorbance of X & Y. Two equations are constructed using this fact.

At λ_1, $A_1 = ax_1\, bc_x + ay_1\, bc_y$ .....(1)

At λ_2, $A_2 = ax_2\, bc_x + ay_2\, bc_y$ .....(2)

      Where,

The absorptivity of X at λ_1 is ax_1, and the absorptivity of X at λ_2 is ax_2

The absorptivity of Y at λ_1 is ay_1 and at λ_2 is ay_2.

The absorbance of the diluted sample at λ_1 and λ_2, A_1 and A_2 respectively.

The concentration of X & Y are C_x & C_y respectively, in the diluted sample.

Rearrange eq. (2).

$$c_y = A_2 - ax_2cx \,/\, ay_2$$

Substituting for c_y in eq. (1). and rearranging gives

$$C_x = A_2ay_1 - A_1ay_2 \,/\, ax_2ay_1 - ax_1ay_2$$
$$C_y = A_1ax_2 - A_2ax_1 \,/\, ax_2ay_1 - ax_1ay_2$$

3. **Absorbance ratio method:** The absorbance ratio method is a modification of the simultaneous equations method. It depends on the property that, for a substance, which obeys Beer's law at all wavelengths, the ratio of absorbances at any two wavelengths is a constant value and is independent of concentration or path length.

In the quantitative estimation of two components in a mixture by the absorbance ratio method, absorbance is measured at two wavelengths, λ_{max} one of the components (λ_2) and wavelength of equal absorptivity of two components (λ_1), i.e. an iso-absorptive point.

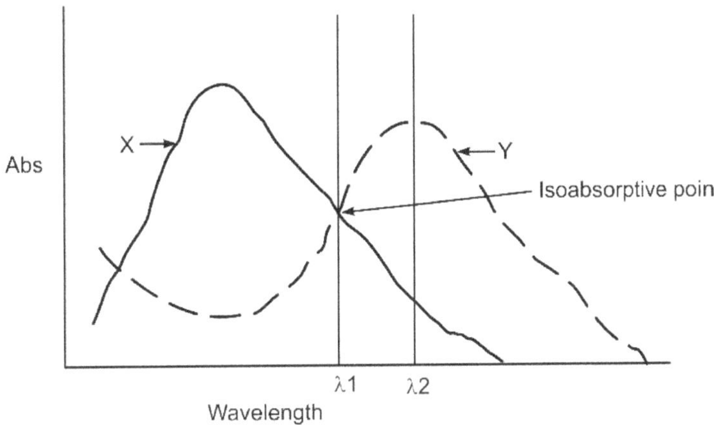

Fig. 1.36 Absorbance ratio method of Atrovastatin and Niacin in tablet dosage forms.

The concentration of the components x and y can be determined by the equation

$$C_X = (Q_m\text{-}Q_y).\, A_1 \,/\, (Q_x\text{-}Q_y).\, ax_1$$
$$C_Y = (Q_m\text{-}Q_x).\, A_1 \,/\, (Q_y\text{-}Q_x).\, ay_1$$

Where $Q_m = A_2 / A_1$ $Q_x = ax_2 / ax_1$ $Q_y = ay_2/ay_1$

A_2 = Absorbance at λ_2; A_1 = Absorbance at λ_1

ax_1 = Absorptivity of Drug X at λ_1

ay_1 = Absorptivity of Drug Y at λ_1

ax_2 = Absorptivity of Drug X at λ_2

ay_2 = Absorptivity of Drug Y at λ_2

E.g., Spectrophotometric method for absorbance ratio method of Atorvastatin and Niacin in the tablet dosage form. λ_{max}: ATR: 246nm NIA: 262nm. The iso-absorptive point at 258nm in Methanol.

4. **Derivative spectrophotometric method:** When the mixture has got interfering absorption, the derivative spectroscopy method is employed.

   Derivative spectroscopy involves the conversion of a normal spectrum to its first, second or higher derivative spectrum.

   The normal spectrum is known as fundamental, zero-order, or D^0 spectra. The plot of the rate of change of absorbance with wavelength against wavelength is called the first derivative spectrum (D1)

   i.e. plot of $\Delta A/\Delta \lambda$ vs λ

   The second derivative spectrum is a plot of $\Delta 2A/\Delta \lambda 2$ vs λ

   The quantitative estimations in derivative spectroscopy give accurate results with second derivative spectroscopy.

   For the quantitative estimation of binary mixtures(A&B) using the derivative spectroscopy, the Zero Crossing Points (ZCP) for both the components (A and B) are noted.

   The ZCP for A and B is selected in such a way that at that particular ZCP, the other component shows remarkable absorbance.

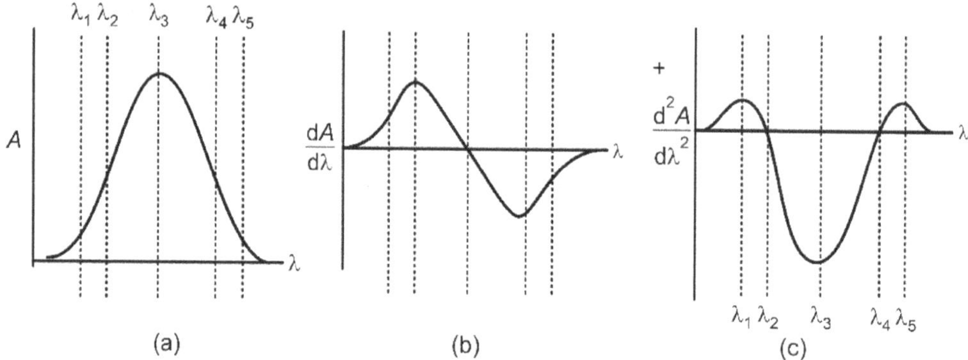

Fig. 1.37 Zeroth (a), First (b), Second (c) derivative spectra

The calibration curve of A is constructed at the ZCP of B and B at the ZCP of A. Unknown concentrations can be found from the calibration graph by extrapolation.

E.g., Simultaneous Spectrophotometric estimation of Ofloxacin and Satranidazole in the tablet dosage form.

5. **Multi wavelength UV-spectrophotometry:** This method helps in the determination of the composition of a binary mixture with overlapping spectra without determining molar absorptivities. The method requires standard solution absorbance of each component and the unknown mixture absorbance

All the standard solutions can be scanned over the range of 200-400nm in the multi-component mode, using two sampling wavelengths. The overlay spectra of standard solutions can be used to determine the concentrations of the drugs in a sample solution.

E.g., Simultaneous Estimation & Validation of Paracetamol, Phenylephrine Hydrochloride and Chlorpheniramine Maleate in Tablets by Spectrophotometric Method:

6. **Dual wavelength method:** The dual-wavelength method is also known as the two wavelengths method. In this method, one of the drugs is considered as analyte and the other drug is considered as an interfering component and vice-versa. The two wavelengths are selected in such a way, that in one wavelength, the interfering component shows the same absorbance (ΔA equals zero), and the analyte shows a significant difference in absorbance with a concentration in the other wavelength. The difference in absorbance (ΔA) between two points on the mixture sample spectra is directly proportional to the concentration of the analyte independent of the interfering component

E.g., Simultaneous determination of atenolol and indapamide

7. **Difference spectroscopy:** Quantitative estimation of the substance in the presence of absorbing interferents can be carried out by this method. The essential feature of this method is that the measured value is the absorbance difference (ΔA) between two equimolar solutions in the same wavelength.

Two equimolar solutions of the analyte in different chemical forms will exhibit different spectral characteristics. The difference in absorption in the selected wavelength is measured.

The alteration in the spectral properties of the analyte is mainly done by the adjustment of the pH using acids, alkalies, or buffers.

Eg: Difference in absorbance of Ibuprofen in 0.1N HCl and 0.1 N NaOH is used in determination of ibuprofen

8. **Geometric correction method:** It is the three-point geometric procedure, the simplest of mathematical correction procedures and it is applied if the interfering absorption is linear at the three wavelengths selected.

If the wavelengths λ_1, λ_2 and λ_3 are selected to that the background absorbances B_1, B_2 and B_3 are linear, then the corrected absorbance D of the drug may be calculated from the three absorbances A_1, A_2 and A_3 of the sample solution at λ_1, λ_2 and λ_3 respectively as follows

The general formula is

$$D = y(A_2 - A_3) + z(A_2 - A_1) / y (1 - w) + z(1 - v)$$

Where

A_1, A_2 and A_3 – three absorbances of the sample

y and z – wave length intervals ($\lambda_2 - \lambda_1$) and ($\lambda_3 - \lambda_2$)

v and w - absorbance ratios vD/D and wD/D respectively

Let vD and wD be the absorbance of the drug alone in the sample solution at λ_1 and λ_3 respectively,

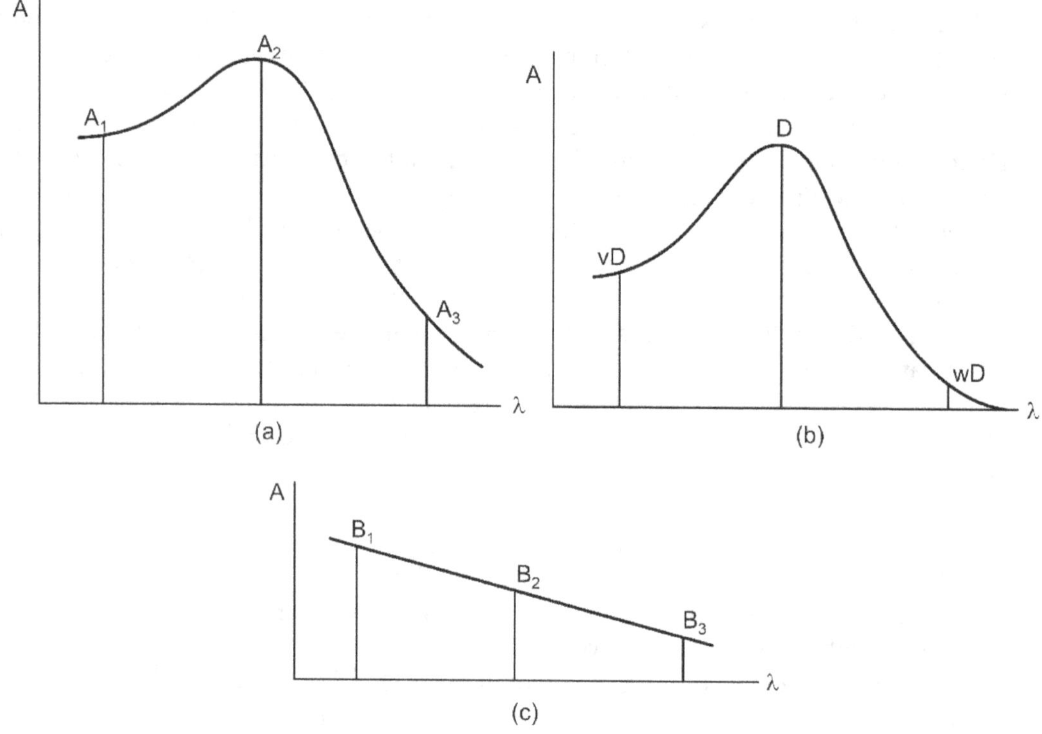

Fig. 1.38 An absorption spectrum (a), the spectrum of analyte (b), background absorption (c).

9. **Orthogonal polynomial method:** It is another mathematical correction procedure, which involves more complex calculations.

   In this method, an absorption spectrum is represented in terms of orthogonal functions like

   $$A(\lambda) = p\, P(\lambda) + p1\, P1(\lambda) + p2\, P2(\lambda) \ldots pn\, Pn(\lambda)$$

   Where A denotes the absorbance at wavelength λ. A set of n+1 equally spaced wavelengths are used at which the orthogonal polynomials, $P(\lambda)$, $P1(\lambda)$, $P2(\lambda)$ ..... $Pn(\lambda)$ are defined

10. **Solvent extraction method:** In this method, different solvents are selected based on the solubility of the components of the mixture and the components are extracted individually. Then the extracted components are analyzed separately.

11. **H-point standard addition method:** The H-point standard addition method (HPSAM) is a modification of the standard addition method. The method is used for resolving strongly overlapping spectra of two analytes and errors produced by the matrix of the sample can also be corrected directly. The technique is a combination of dual-wavelength spectrophotometry and the standard addition method.

12. **Least square approximation:** Experimental measurements are always subject to random errors. To correct these errors and to obtain an accurate result, more number of experimental data is used. That is, the analysis of a binary mixture is carried out in three or four wavelengths instead of at two wavelengths,

The method of least squares is a standard approach to the approximate solution of overdetermined systems, i.e., sets of equations in which there are more equations than unknowns. "Least squares" means that the overall solution minimizes the sum of the squares of the errors made in the results of every single equation.

Other Applications of UV spectroscopy

1. **Qualitative analysis:** Identification of the compounds is done by comparing the absorption spectrum with the spectra of standard known compounds.

2. **Detection of Impurities:** UV spectroscopy is one of the best methods for the detection of impurities in organic compounds. When comparing the absorption spectrum of the sample with a standard absorption spectrum, the presence of additional peaks indicates the presence of impurities. Impurities can also be detected by measuring in specific wavelengths characteristic for them

   E.g., The impurity, cyclohexane, in benzene can be detected by measuring the absorption at 255nm.

3. **Structural elucidation of Organic compounds:** Structural elucidation like the presence/absence of unsaturation, presence of heteroatoms like S, O, N, or halogens can be determined using UV spectroscopy. The general absorptions of these features are given

   E.g. Saturated compound $\sigma - \sigma^*$ appears below 200nm (vacuum uv)

   The other three possibilities are given below, and they appear between 200-400 nm.

   Presence of heteroatoms (S, N, O) with saturation $n - \sigma^*$

   Unsaturated compounds $\pi - \pi^*$

   Hetero atom and double bond $n-\pi^*$

   Thus the location of peaks in the absorption spectrum gives an idea about the structural features.

4. **Structural analysis of organic compounds:** The effect of conjugation, cross conjugation, and alkyl substitution are discussed here

   (i) **Detection of conjugation:** It helps to know the presence of different groups with respect to conjugation, e.g., The conjugation maybe

   > Between two or more C=C (or triple) bonds

   > Between C=C and C=O

   > Between double bonds and aromatic rings.

   The presence of aromatic rings can be identified and the no and positions of attachment to the carbons of the conjugated system can also be identified.

   Extension of the conjugation gives **bathochromic** shift or redshift by shifting the λ_{max} to longer wavelength.

   Reduction or saturation of double bonds leads to **hypsochromic** shift or blue shift by shifting the λ_{max} to the shorter wavelength.

(ii) Detection of Geometrical Isomerism: The **trans isomer** of the geometrical isomers absorb at a **longer wavelength** and have a **larger extinction coefficient** than cis isomers.

Eg: Calciferol- λ_{max} 265 nm (cis)

Iso vitD$_2$-λ_{max} 287 nm (trans)

Conversion of Cis – trans-isomer shows Bathochromic shift & Hyperchromic effect

Conversion of Trans to cis exhibits Hyposochromic shift and Hypochromic effect

(iii) Cross conjugation: The cross conjugation has got no effect on λ_{max}.

(iv) Alkyl substitution: Bathochromic shift occurs with alkyl substitution. More the no of alkyl substitution more is the shift.

(v) No of rings: Bathochromic shift occurs with the addition of rings.

CHAPTER 2

Fluorimetry

INTRODUCTION

A large number of substances absorb ultraviolet or visible radiation. Some of these substances lose a part of their excess energy as electromagnetic radiation in a longer wavelength than the absorbed wavelength. This process of emitting radiation is known as **luminescence**. In luminescence, the light is produced at low temperatures. The types of luminescence are

(a) Photoluminescence
(b) Chemiluminescence
(c) Bio luminescence
(d) Electroluminescence
(e) Radio luminescence
(f) Thermo luminescence.

Photoluminescence is caused by moving electrons to energetically higher levels through the absorption of photons. It is further subdivided into

(i) **Fluorescence** - When a beam of light is incident on certain substances, they **emit visible light or radiations**. This phenomenon is known as **fluorescence** and the substances showing this phenomenon are called **Fluorescent substances**.

Fluorescence is **instantaneous** and starts immediately after the absorption of light and stops as soon as the incident light is cut off.

Materials exhibiting fluorescence emit radiation within 10^{-6} **to** 10^{-4} seconds of absorption.

(ii) **Phosphorescence** – When light radiation is incident on certain substances, they emit radiation continuously **even after the incident light is cut off.** This type of **delayed fluorescence** is called phosphorescence. Substances showing this phenomenon are called **phosphorescent substances**.

Materials exhibiting phosphorescence emit radiation within 10^{-4} **to 20** seconds or more.

The lifetime of phosphorescence is much longer than fluorescence.

Most of the molecules in their ground state do not have unpaired electrons (singlet state). When such a molecule absorbs UV or Visible radiation, the electrons (π electrons) get raised to an excited singlet state where there is no change in the spin of the electrons. If one set of electron spins undergo unpairing (change in spin), the resulting electrons will make an excited Triplet state.

From the excited singlet state, the molecule will return to the ground state by any one of the following possibilities

1. Returning to the ground state by **collisional deactivation** without emitting any radiation.
2. The molecule in the excited singlet state may **emit UV or Visible light** photons. This is known as **fluorescence.**
3. The molecule in the excited singlet state may undergo transition to **metastable triplet state (intersystem crossing)** and then return to the ground state by **emitting UV or Visible light** photons. This is known as **phosphorescence.**

Principle

International Union of Pure and Applied Chemistry (IUPAC) define fluorescence as spontaneous emission of radiation (luminescence) from an excited molecular entity with the formation of a molecular entity of the same spin multiplicity.

When a beam of light is incident upon certain substances, they emit visible light or UV radiation. This phenomenon is called **fluorescence** and the substances exhibiting this property are called Fluorescent substances. It is instantaneous. It starts as soon as the substance absorbs the light and stops immediately after the incident radiation is cut off.

Absorption of UV or visible radiation causes the transition of electrons in the molecules from the singlet ground state to the excited singlet state. As the excited state is not stable, the electrons return to the ground state by losing energy in the form of UV/visible radiation. Thus the principle of fluorimetry is the study or measurement of the emitted radiation when the electrons return from the **excited singlet state to ground singlet state** after absorbing light.

Electronic states

The different electronic states are

1. Singlet ground state
2. Doublet state
3. Triplet state
4. Singlet excited state.

These terms tell us the **number of unpaired electrons in the absence of a magnetic field.** There will be n+1 number of degeneracy states (equal energy) associated with the electron spin if there are n number of unpaired electrons. For no unpaired electrons (n=0), there is only one spin state. (n+1 =1).This state is called as singlet state. Thus,

Singlet ground state is a state where there are no unpaired electrons in a molecule. i.e. all electrons are paired. ↑↓

In **Doublet state** one unpaired electron is present (n + 1 = 1 + 1 = 2) ↑ or ↓

In the **Triplet state,** two unpaired electrons are present. Both electrons have the same spin

↑↑

In the **singlet excited state** also there are **no unpaired electrons.** Though the electrons are in an excited state still they are paired **(No change in spin)**

Excitation states

Pauli's Exclusion Principle states that no two electron in an atom can have the same set of 4 quantum numbers, i.e. principal, azimuthal, magnetic and spin quantum number. This restriction requires that no more than 2 electrons can occupy an orbital and further that the two must-have opposite spin states. Under these circumstances, the spin is said to be paired. Because of spin pairing, most molecules exhibit no net magnetic field and are thus said to be diamagnetic, i.e. they are neither attracted nor repelled by the static magnetic field.

Singlet excited state. A molecular electronic state in which all electron spin are paired even in the excited state is called a singlet excited state. No splitting of electronic levels occurs when the molecules are exposed to a magnetic field. Electrons are unpaired, but the **net spin is zero**.

A Triplet state is a state when one of the paired electrons of the molecule is excited to a higher energy level, it forms an excited singlet state, and when the spin of the excited electron changes and becomes unpaired, the state is termed as triplet state. The property of molecule differs in triplet state than in the singlet-excited state, for example, paramagnetic and diamagnetic. Singlet excited to triplet transition, which involves a change in the spin of electron, is significantly less probable event than singlet ground to singlet excited state transition.

Singlet excited state is the state where one of the paired electrons absorbs light and goes to the excited state resulting in unpairing, but there is no change in the spin of the electron.

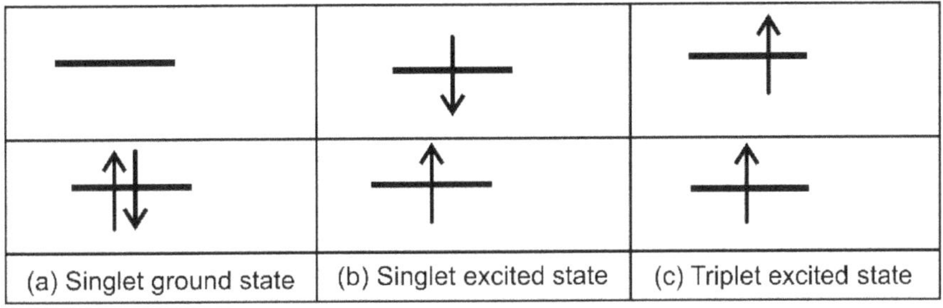

| (a) Singlet ground state | (b) Singlet excited state | (c) Triplet excited state |

Fig. 2.1 Different electronic states of a molecule after the absorption of light.

Fluorescence process can be broken down into three phases

1. **Excitation** – Excitation is the absorption of light of an appropriate wavelength by a fluorophore and the transition of an electron from ground level to an excited level. Chromophores which exhibit the phenomenon of fluorescence are called fluors or fluorophores. Fluorophores are organic molecules of 20-100 Daltons.

2. **Excited-state** – After the absorption of a photon of appropriate energy, fluorophore undergoes vibrational and conformational changes

3. **Emission** – Release of excess energy in the form of a photon of light is emission phase. The fluorescence process is cyclical; therefore, a fluorophore can be excited repeatedly. Absorption of UV or visible radiation causes transition of electrons in the molecules from the singlet ground state to the excited singlet state. As the excited state is not stable, the electrons return to the ground state by losing energy

   Loss of energy can occur by these ways:
   1. **Vibrational relaxation/ Collisional deactivation**
   2. **Internal conversion**
   3. **External conversion**
   4. **Intersystem crossing**

5. Fluorescence

6. Phosphorescence

1. Vibrational relaxation/Collisional deactivation: Vibrational energy is the energy used to lengthen or shorten bonds or widen or squeeze bond angles. An electron in a molecule may be promoted to any of the several vibrational levels during the electronic excitation process. In solution, the excess of vibrational energy is immediately lost as a consequence of the collision between the molecule of the excited species and those of solvent. The result is an energy transfer and a minuscule increase in the temperature of the solvent. This relaxation process is so efficient that the average lifetime of a vibrationally excited molecule is 10-12 sec or less (a period significantly shorter than the average lifetime of an electronically excited state. As a result, fluorescence from solution when it occurs involves a transition from the lowest vibrational level of an excited state). This also results in the displacement of fluorescence band towards a longer wavelength than the absorbed one; this is called **Stokes shift.**

Vibration relaxation of an excited electron from the highest vibrational level in singlet excited state to lowest vibrational level results in stokes shift which is shown in figure.

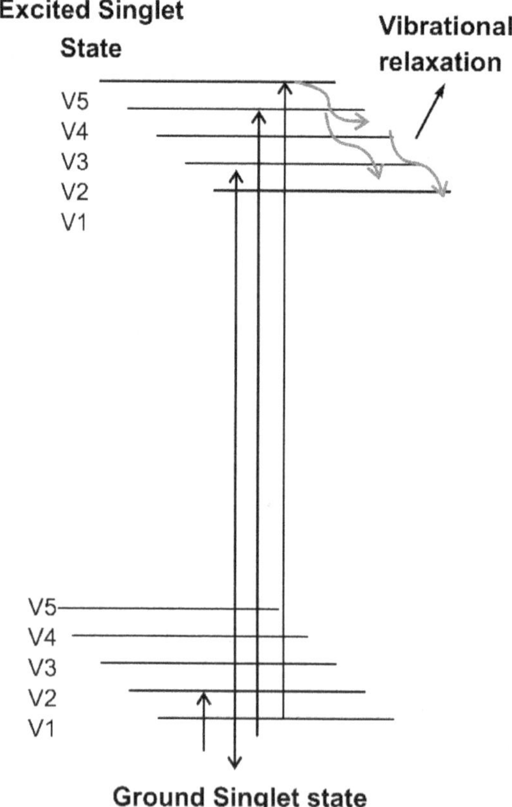

Fig. 2.2 Vibrational relaxation.

2. **Internal conversion (Radiationless Transitions):** Electrons tend to avoid giving off a giant photon if they can get to a lower energy state by giving off small packets of energy at a time, by simple hopping down to lower and lower vibrational levels. They may relax all the way down to the ground state via vibrational relaxation if any of the excited state's lower vibrational levels have energy that overlaps with some of the higher vibrational levels of the ground state.

Internal conversion is a process by which a molecule passes to a lower electronic state without the emission of radiation. Internal conversions are efficient when two electronic energy levels are significantly close for the existence of overlap in vibrational energy level, i.e. the potential energies of the two excited states are identical. This equality permits efficient transition through overlapping, which is usually more probable than the loss of energy by fluorescence from a higher excited state.

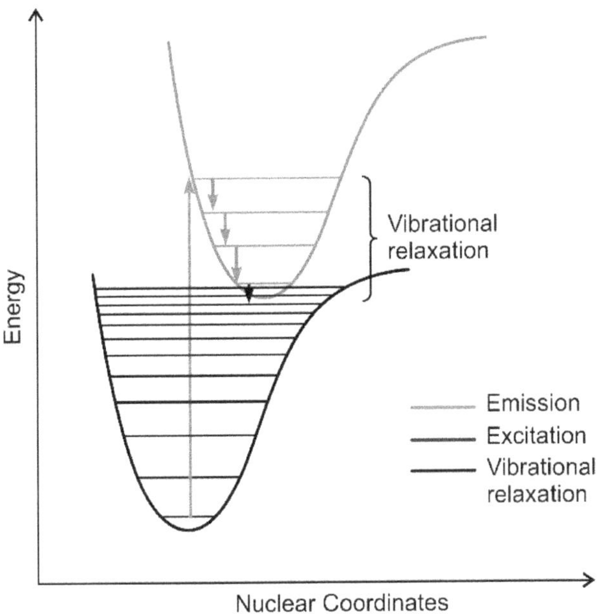

Fig. 2.3 Internal conversion.

3. **External conversion:** External conversion involves interaction and energy transfer between the excited molecule and the solvent or other solute. This markedly affects the fluorescence intensity. Conditions which tend to reduce the number of collisions between particles (low temperature, high viscosity) generally lead to enhanced fluorescence. Radiationless transition to ground state from the lowest excited singlet state and triplet state probably involve external conversion as well as internal conversion.

4. **Intersystem crossing:** An organic molecule with no unpaired electrons in its ground state is said to be in a singlet state. According to one of the selection rules for electronic excitation, the excited state must also have no unpaired electrons. In other words, the

spin of the electron that gets excited is the same after excitation as it was before excitation.

The intersystem crossing is a process in which there is a change in multiplicity of the molecule, and the spin of an excited electron is reversed from + 1/2 to -1/2 or vice versa. This results in an electronic state of lower energy (single electrons in different orbitals prefer to have parallel spin) with a multiplicity (possible orientations of the spin) M= 3, which is thus designated as a triplet state, T1. (multiplicity is $2S+1 = 3$, so that the total spin $S = 1$. This spin is due to two unpaired electrons based on Hund's rule, favouring degenerate orbitals' single filling. The triplet consists of three states with spin components +1, 0 and –1). This is enhanced if the vibrational level of the two states overlaps. The intersystem crossing is most common in the molecule that contains heavy atoms, i.e., iodine, bromine and is also known as the heavy atom effect. The interaction between the spin and orbital motion becomes large in the presence of such an atom. The presence of paramagnetic species such as molecular oxygen in solution also enhances intersystem crossing and as a result, decreases fluorescence.

5. **Fluorescence**: The molecule in the excited state emits UV or visible radiation of a longer wavelength to lose energy. **The emission of radiation occurs at a longer wavelength** because the energy of emitted radiation is lesser than the energy of absorbed radiation. This happens because some amount of the absorbed energy is lost by vibrational or collisional deactivation.

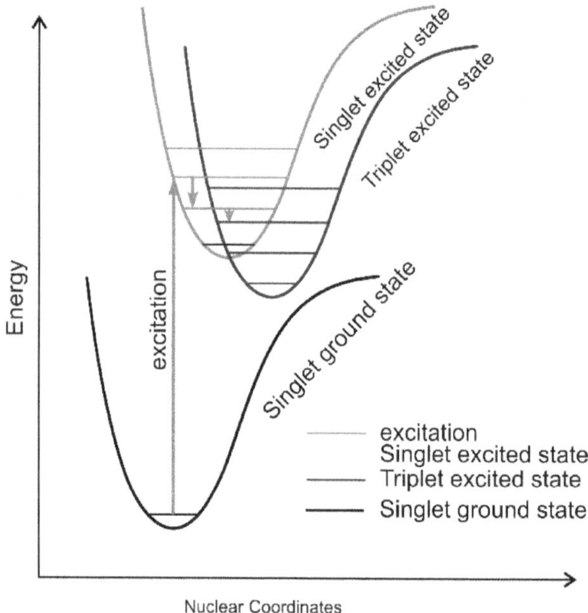

Fig. 2.4 Intersystem crossing.

Thus, fluorescence is the phenomenon of emission of radiation from singlet excited state to singlet ground state. The absorption wavelength is called excitation wavelength, and that of the emitted radiation is called emission wavelength. These two wavelengths are characteristic for a given substance under given conditions.

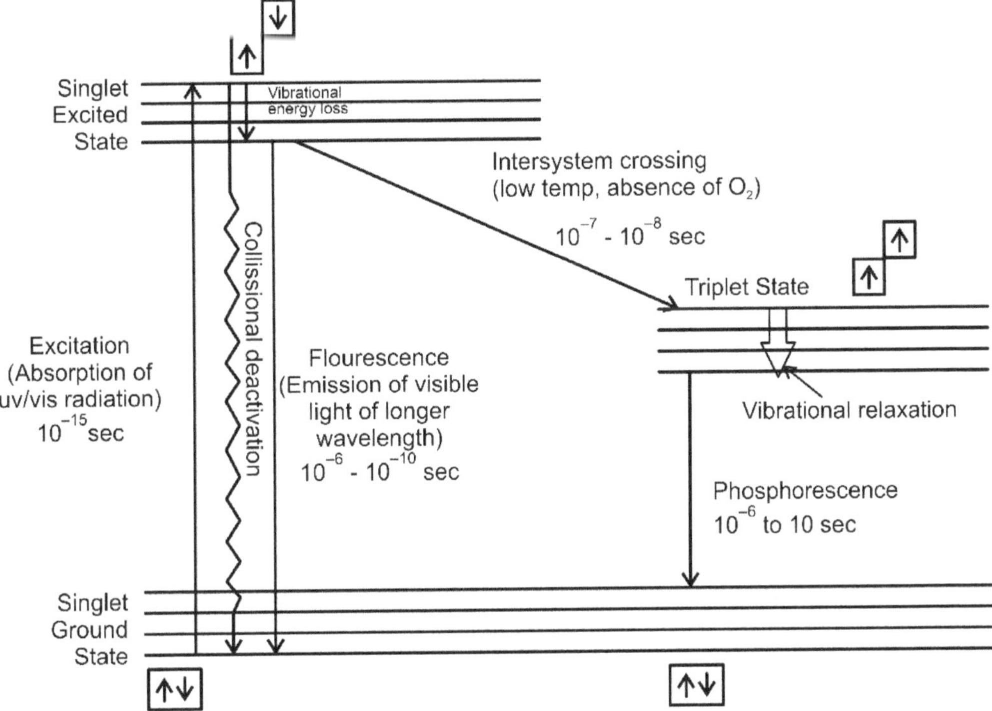

Fig. 2.5 Jablonski diagram.

6. **Phosphorescence:** The molecule in the unstable excited singlet state may undergo transition to a metastable triplet state and from there returns to the ground state by emitting UV/Visible radiation. This is called phosphorescence. **The process of transition from excited singlet state to excited triplet state is called intersystem crossing**. It occurs at favourable conditions like **low temperature** and the **absence of oxygen**. The energy of the excited triplet state is lower than the excited singlet state, and the emission occurs at a longer wavelength. The lifetime of phosphorescence is longer, and it continues even after the incident radiation is cut off. It is also called **delayed fluorescence**.

FACTORS AFFECTING FLUORESCENCE

The factors affecting fluorescence can be classified as two types

1. Structural factors
2. Non-structural factors

1. Structural factors

   (i)   Nature of molecules
   (ii) Nature of substituent groups

(iii) Rigidity of structures

(iv) Effect of pH

2. Non -structural factors

(i) Effect of concentration

(ii) Adsorption

(iii) Light

(iv) Effect of Temperature and viscosity

(v) Effect of oxygen

(vi) Photo-decomposition

(i) Nature of molecules: All molecules which absorb radiation cannot show the phenomenon of fluorescence and phosphorescence. The molecules must possess conjugated double bonds (π electrons) so that the UV/visible radiation can be absorbed and fluorescence can be observed.

(ii) Nature of substituent groups: Electron donating groups like amino (NH_2) and hydroxyl (OH) groups enhance fluorescence intensity. Groups like SO_3H, NH_4^+ and alkyl groups do not have much effect on fluorescence and phosphorescence.

Electron withdrawing groups like -COOH,-NO_2,-N=N- and halides reduce or sometimes even destroy the fluorescence intensity.

If an atom with a high atomic number is introduced in the π electron system, this enhances phosphorescence and decreases the Fluorescence intensity.

Table 2.1 Effect of substituent on fluorescence intensity.

Substituent	Effect on λ	Effect on intensity
Alkyl	Not affected	Not much effect
COOH,CHO,COOR,COR	Shifts to Longer	Decrease in intensity
OH, OMe, OEt	Shifts to Longer	Increase in intensity
CN	Not affected	Increase in intensity
NH_2, NHR, NR_2	Shifts to Longer	Increase in intensity
NO_2, NO	Shifts to Longer	Quenching
SH	Shifts to Longer	Decrease in intensity
SO_3H	Not affected	Not affected
F, Cl, Br, I	Shifts to Longer	Decrease in intensity

(iii) Effect of concentration: Fluorescence **is proportional to concentration,** but this relationship holds only when the **concentration of the substance is low** (μg or ng/ml). **Higher concentration** leads to a **decrease** in fluorescence intensity

(iv) Adsorption: Only dilute solutions should be used in fluorimetry. Concentrated solutions lead to the adsorption of fluorescent substances on the container walls of sample cell leading to decreased fluorescence. E.g., Quinine.

(v) Light: Monochromatic light is essential for quantitative fluorescent analysis because the intensity varies with wavelength.

(vi) Rigidity of structures: Rigid structures give **more** fluorescence intensity, and the **flexible structures** provide **less intensity.**

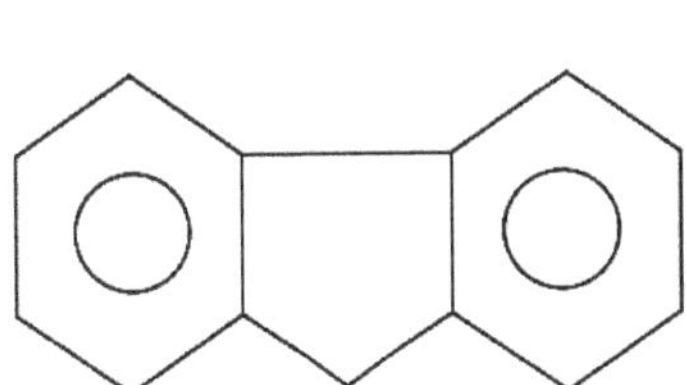

Fig. 2.6 Rigid structure Fluorene   Fig. 2.7 Flexible structure Biphenyl

(vii) Effect of Temperature and viscosity: An increase in temperature leads to a decrease in viscosity which results in decreased fluorescence. This is because the excited molecules get deactivated by collision. A decrease in temperature gives increased fluorescence due to decreased collisions (increase in solvent viscosity).

(viii) Effect of Oxygen: The presence of oxygen decreases the fluorescence intensity. It can happen in two ways

   1. It can oxidize the fluorescent substance to a non-fluorescent substance.
   2. Because of its paramagnetic properties (triplet ground state), oxygen quenches fluorescence.

(ix) Effect of pH: A change in pH will change the ionic state of the molecule. Thus, the pH effect is dependent on the chemical nature of the molecule.

   E.g.,

   1. Aniline in an acid medium (ionized) gives fluorescence in the UV region, whereas in a neutral and alkaline medium (unionized) gives visible fluorescence.
   2. Phenols are undissociated in acid medium and don't exhibit fluorescence, but in the alkaline condition, they are in an ionic state giving fluorescence.

(x) Photodecomposition: The magnitude of photodecomposition depends upon the intensity of the light source. Spectrofluorimetry requires a high intensity of radiation to enhance the sensitivity of the instrument, thus leading to photochemical decomposition. Hence the wavelength which is not strongly absorbed should be chosen to eliminate such decompositions. If not error of upto 20% is possible in the measurements.

TYPES OF FLUORESCENCE

Fluorescence and phosphorescence come under the classification of photoluminescence.

Based on the emitted radiation wavelength, fluorescence can be classified as

   (a) Stokes fluorescence: In this type of fluorescence, the wavelength of the emitted radiation occurs at a longer wavelength. This is the conventional fluorimetric experiment.

(b) Anti-stoke's fluorescence: The wavelength of emitted radiation is shorter than the absorbed radiation. This type of fluorescence occurs in thermally assisted fluorescence.

(c) Resonance fluorescence: This is the fluorescence in which both the emitted radiation and the absorbed radiation are equal. E.g., Mercury vapour at 254 nm.

QUENCHING AND ITS TYPES

QUENCHING

Quenching is any process that reduces the fluorescence intensity of a given substance.

This may occur due to various factors like pH, concentration, temperature, viscosity, presence of oxygen, heavy metals, specific chemical substances, etc. Example of **quenching agents: Thiocyanate, Chloride, Iodide, Disulfide, Nitric oxide.**

TYPES OF QUENCHING

1. Collisional Quenching
2. Static Quenching
3. Chemical Quenching
4. Concentration quenching/self-quenching

1. **Collisional Quenching:** Collisional quenching occurs by the increased collision of quencher molecules with excited molecules of the fluorescing substance. Halides ions such as chlorides or iodides, increased temperature and decreased viscosity are well-known collisional quenchers. For example, quenching of quinine drug by chloride ion or tryptophan by iodide ion follow the collisional quenching process.

2. **Static Quenching:** A complex formation occurs between the fluorescing molecule at the ground state and the quencher molecule through a strong coupling. Such complex may not undergo excitation or may be excited to a little extent reducing the fluorescence intensity of the molecule. Caffeine and related xanthenes and purines reduce the intensity of riboflavin by static mechanism. Quenching that occurs due to oxygen also follows this mechanism.

3. **Chemical Quenching:** Chemical quenching is due to various factors like - change in pH, presence of oxygen, halides, electron-withdrawing groups and heavy metals, etc.

   (a) Change in pH: Aniline at pH (5-13) gives blue fluorescence when excited at 290 nm. But in pH <5(occurs as a cation) or pH >13(occurs as an anion) does not show any fluorescence.

   (b) Oxygen: Oxygen leads to the oxidation of fluorescent substances to non-fluorescent substances and thus causes quenching due to its paramagnetic property (triplet ground state).

   (c) Halides and electron-withdrawing groups: Halides like chloride ions, iodide ions and electron-withdrawing groups like -NO, -COOH, -CHO groups lead to quenching.

   (d) Heavy metals: The presence of heavy metals also leads to quenching because of collision and complex formation.

4. **Concentration quenching / self-quenching:** Concentration quenching is a kind of self-quenching. It occurs when the concentration of the fluorescing molecule increases in a sample solution. The fluorescence intensity is reduced in a highly concentrated solution (>50 µg/ml).At high concentrations, there is no proportional increase in fluorescence intensity with an increase in concentration. There is a decrease in fluorescence intensity with an increase in concentration. This phenomenon is called concentration quenching or self-quenching.

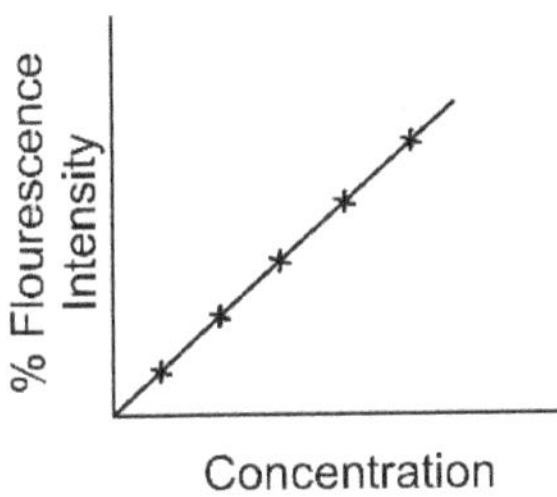

<table>
<tr><td>Fig. 2.8(a) Calibration curve in low concentration</td><td>Fig. 2.8(b) Calibration curve in high concentration</td></tr>
</table>

At higher concentrations, there is a deviation from linearity. i.e., there is no proportional increase in fluorescence with an increase in concentration, but there is a decrease in fluorescence. **This is called self-quenching or concentration quenching.** This phenomenon occurs because of the internal circulation of energy(radiation) caused by the absorption and re-emission of radiation by adjacent molecules before falling on the detector. The concentration used for the fluorescence measurement should be in the beer's law range.

Effect of concentration on fluorescence intensity

$$\textbf{Fluorescence intensity} = \textbf{Q} \times \textbf{I}_a$$

where Q- fluorescence efficiency

$$\text{Fluorescence efficiency} = \frac{\text{Fluorescence intensity emitted}}{\text{Electromagnetic radiation intensity absorbed}}$$

I_a = Intensity of absorbed light

$I_a = I_o - I_t$ since emission is proportional to absorption. .....(2.1)

where I_o = intensity of incident light and

I_t = intensity of transmitted light

By Beer-Lambert's law $\quad I_t = I_0 \, e^{-act}$

Substituting this for I_t in equation 1

$$I_a = I_o - I_0 \, e^{-act}$$
$$I_a = I_o \, (1 - e^{-act})$$
$$I_a = I_o \, [1 - (1 - act)]$$
$$I_a = I_o \, [1 - 1 + act]$$
$$I_a = I_o \times act$$

Therefore, Fluorescence intensity $= Q \times I_a = QI_0act$

i.e., $\qquad$ $F = QI_0act$

where $\qquad$ $Q =$ constant for the given substance

$\qquad\qquad$ $I_0 =$ constant for the instrument

$\qquad\qquad$ $a =$ molar extinction coefficient and it is constant for the substance

$\qquad\qquad$ $t =$ pathlength (constant for the given cell)

$\qquad\qquad$ $c =$ Concentration of the substance

$\therefore$ $\qquad\qquad$ $F = KC$ where $K = QI_0a$ and t

Or in other words, **fluorescence is directly proportional to Concentration.**

But this relationship holds good only in dilute solutions in the range of µg or ng/ml.

In high concentrations (mg/ml), it does not obey beers law (No linearity observed).

INSTRUMENTATION

The instruments used for fluorimetric analysis are called **fluorimeters or fluoro photometers.** In these instruments, **filters** are used for spectral wavelength isolation.

Spectrofluorimeters employ **two monochromators** to isolate the required wavelength. These instruments are used for scanning the spectrum of wavelength continuously and automatically.

Table 2.2 Summary of instrument components.

Light Source	Filters and Monochromators	Sample cells	Detectors
Mercury vapour lamp Xenon arc lamp Tungsten lamp	Primary filter secondary filter excitation and emission monochromators (gratings)	Cylindrical glass cuvettes Pathlength 1cm All sides polished	Photomultiplier tubes, Photovoltaic cell, Photo tubes placed at 90° angle to incident radiation

CONSTRUCTION OF FLUORIMETERS

1. **Light Source:** The light source should emit radiation continuously over the required wavelength region. It should be stable and provide intense radiation.

    (a) **Mercury vapour lamp:** High pressure (8 atmospheres) mercury vapour gives intense lines in a continuous background above 350 nm. Lines are obtained at 365, 398, 436, 546, 579, 690 and 734 nm. An additional line at 254 nm is obtained with a low-pressure mercury lamp which is used as a source in filter type fluorimeters

    (b) **Xenon arc lamp:** The radiation is intense when compared to mercury vapour lamps. They are used in Spectrofluorimeters.

    (c) **Tungsten lamp:** This lamp is used if the excitation wavelength is in the visible region. This lamp cannot work in the UV region. The intensity of this lamp is very low.

2. **Filters and Monochromators:** In fluorimetry, the excitation wavelength and the emission wavelengths are different. A filter or monochromator is used for the selection of these wavelengths. In photofluorimeters, **primary filters and secondary filters** are used for this purpose.

    Primary filter absorbs visible radiation and transmits UV radiation.

    Secondary filter absorbs UV radiation and allows the visible radiation to pass through.

    Spectrofluorimeters contain excitation and emission monochromators with gratings.

     Excitation monochromator selectively allows only the wavelength absorbed by the sample (excitation wavelength).

     Emission monochromator isolates and allows only the **radiation emitted** by the fluorescent molecule (emission wavelength).

    Filters and monochromators are explained in UV spectroscopy.

3. **Sample cells:** The sample cells, known as cuvettes, are used to hold the sample.

    The cells are made of colour corrected fused glass.

    The cells are available in different shapes like cylindrical or rectangle.

    Small volume cells (0.5ml or less) and large volume cells (5-10ml) are available depending on their capacity to hold sample volume. The common pathlength (internal distance) is 1cm.Long path length cells with 10cm and short pathlength cells with 1 or 2mm are also available.

    In colorimetry, one set of opposite sides of the cells are polished (for transmission of light), and the other set of opposite sides are ground (for handling).But in fluorimetry, all surfaces are polished so that the **fluorescence measurements** can be made at **90° angle to the incident radiation.**

4. **Detectors:** As the **emitted** radiation is mostly **visible** radiation and sometimes UV radiation, the detectors used are **Photovoltaic cells, Photo tubes, or Photomultiplier tubes.**

    Photomultiplier tubes are more **accurate and best** because when low concentration of the substance is used, only weak radiation is emitted.

Detectors are explained in the UV chapter.

The common types of instruments are

(i)   Single beam filter fluorimeters

(ii) Double beam filter fluorimeters

(iii) Double beam Spectrofluorimeters

(i)   **Single beam filter fluorimeters: These are inexpensive instruments containing tungsten lamps as the source. The optical system consists of a primary filter that absorbs** visible radiation and transmits the UV radiation for the excitation of the molecules of the sample in the sample cell. The emitted fluorescence is measured at 90° using the secondary filter, which absorbs UV radiation and allows the visible radiation to pass through to the detector. If the detector is used at 180° as in colourimetry, both the emitted radiation and the transmitted radiation will produce detector response leading to error in measurements.

Advantages:

1. Simple in construction
2. Cheaper
3. Easy to use

Disadvantages:

1. Sample and reference solutions cannot be measured simultaneously.
2. Rapid scanning is not possible

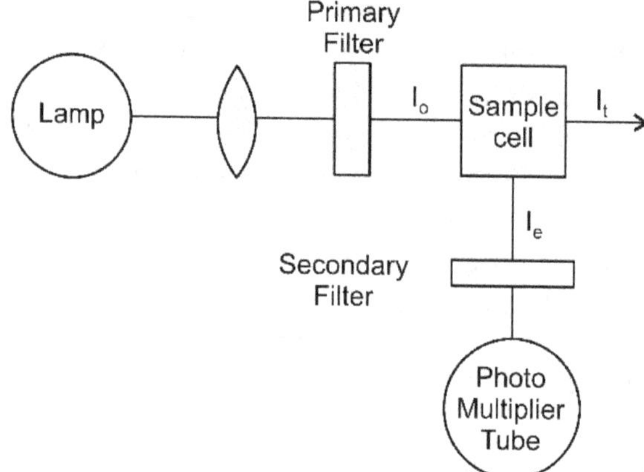

Fig. 2.9 Single beam filter fluorimeter.

(ii) Double beam filter fluorimeters: The construction is the same as a single beam except that from a single light source, two incident beams pass through two primary filters

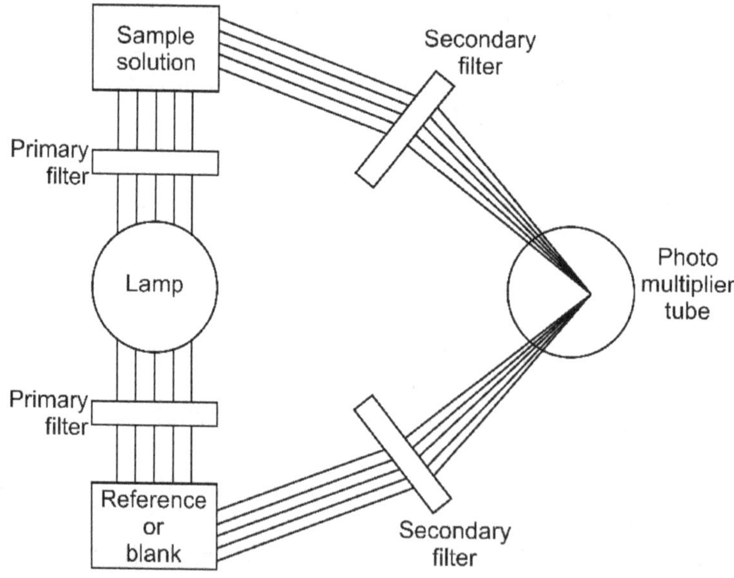

Fig. 2.10 Double beam filter fluorimeter.

separately and fall on the sample and reference standard. The emitted radiation from the sample and reference pass separately through two secondary filters to the detector, and a combined response is obtained from the detector.

Advantage:

Sample and reference can be analyzed simultaneously.

Disadvantage:

Sensitivity and accuracy is less due to the use of filters

(iii) Spectrofluorimeter (Double beam): In Spectrofluorimeters, the excitation and emission monochromators are used instead of primary and secondary filters of double beam filter fluorimeter. A beam splitter is used to split the incident beam into a sample and reference beam. A **Photomultiplier tube** is used as a detector.

Advantages:

1. Rapid scanning

2. More sensitive and accurate.

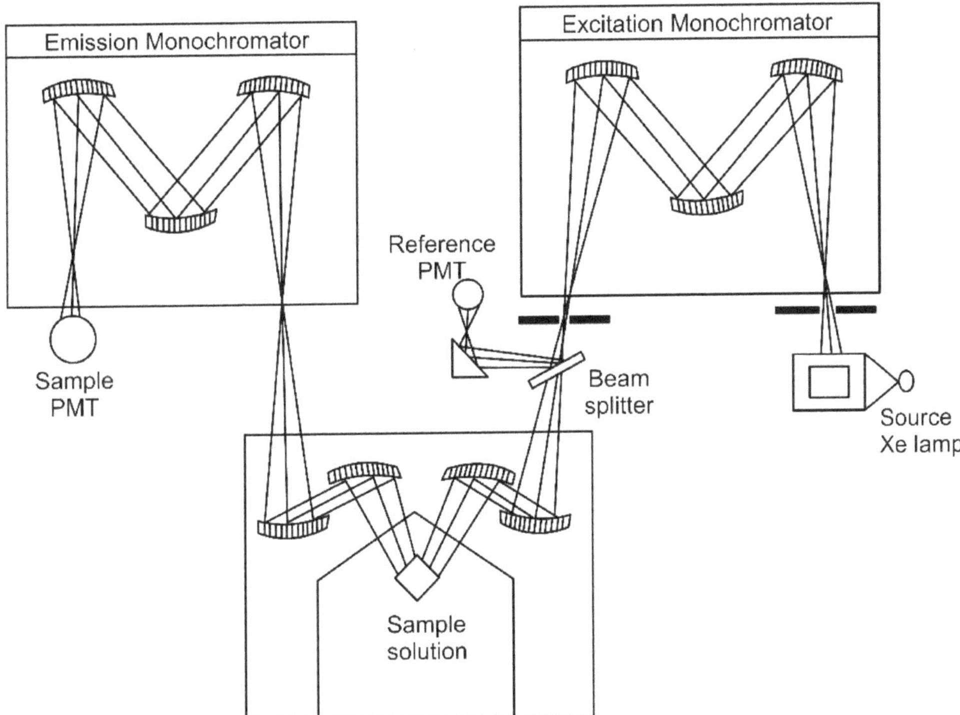

Fig. 2.11 Double beam spectrofluorimeter.

ADVANTAGES OF FLUORIMETRY

1. **Sensitivity** is more compared to other absorption techniques because concentrations like µg or ng /ml can be analyzed by fluorimetry.

2. **Accurate:** The accuracy of fluorescence is approximately 100 times more than that of an absorption procedure.

3. **Specific:** Excitation and the emission wavelengths are characteristic for the substance, making fluorescence more specific than the absorption methods.

4. **Application range:** Using chemical reactions, even non-fluorescent compounds can be converted to fluorescent compounds.

Limitations:

1. **Variation in pH** affects fluorescence intensity.
2. **The presence of oxygen** affects fluorescence intensity.
3. **The presence of Heavy metals, halides** affect fluorescence intensity.
4. All substances are not fluorescent in nature.
5. UV absorption may cause chemical changes in some substances.

APPLICATIONS

1. **Determination of inorganic compounds**
   (a) Generally, inorganic ions do not possess fluorescence. But they form fluorescent chelates with non-fluorescent organic substances.
      (i) E.g., a. Determination of ruthenium ion in the presence of platinum by forming a complex with 5- methyl-1,10-phenanthroline
   (b) Determination of aluminium III in alloys with Pontachrome blue-black RM
   (c) Estimation of traces of boron in steel by complexation with benzoin.
   (d) Quantification of calcium with calcein by fluorimetry.

2. **Fluorescent indicators:** Fluorescent indicators are substances whose **intensity and colour depend on pH range.** These are used in acid-base titrations. Examples with pH range are given below

   E.g., Eosin(3-4), Fluorescein (4-6), Quinine sulphate(3-5), 2-Naphthaquinone (4.4-6.3), Acridine (5.2-6.6)

3. **Fluorometric reagent:** They are used **for cation analysis.** They form chelates with metal ions. E.g.,
   (a) Alizarin garnet B for Al^{3+} determination
   (b) 8-Hydroxy quinoline in Li^+ determination
   (c) Flavanol in Sn^{4+} determination.
   (d) Benzoin in Zn^{2+} determination

4. **Determination of vitamin B**
   (a) **Vitamin B$_1$ (Thiamine),** which is **non-fluorescent,** is converted to **fluorescent thiochrome** (blue fluorescence) by **oxidation** with oxidizing agents like potassium ferricyanide. The fluorescence is measured, which is directly proportional to concentration compounds.
   (b) **Vitamin B$_2$ (Riboflavin)** is determined based on the principle that **riboflavin on oxidation gives a non – fluorescent substance.**

5. **Organic analysis:** Qualitative and quantitative analysis of aromatic compounds present in cigarette smoke, air pollution, and automobile exhausts is done using fluorimetry.

    E.g., indoles, Naphthols, Proteins, plant pigments, and steroids.

6. **Pharmaceutical applications**

    (a) **Direct Fluorescent analysis:** In this method, Fluorescent drugs are determined directly. Eg: Aminacrine, P-amino salicylate, Griseofulvin, Quinine, Riboflavin.

    (b) **Indirect method:** The drugs are converted to fluorescent products by chemical reaction with other reagents.

    E.g., Chlordiazepoxide undergoes Photo-oxidation

    Hydrocortisone treated with 75%v/v sulphuric acid in ethanol

    Nicotinamide treated with cyanogen bromide.

    Diphenyl Hydantoin treated with alkaline Potassium permanganate.

    Thiamine is oxidized with potassium ferricyanide.

    (c) **Analysis of Mixtures**

      (i) **Intensity and colour are** chosen in such a way as to excite only one drug.

      (ii) Compounds can be separated and analyzed.

      (iii) **Using chemical reaction:** Both Codeine and Morphine give fluorescence at 355 nm. But in alkaline conditions, only codeine gives fluorescence.

      (iv) **Derivatization:** Atropine forms a complex with eosin which gives fluorescence when dissolved in chloroform.

CHAPTER 3

IR Spectroscopy

IR spectroscopy is a valuable tool used for the structural determination of organic compounds in conjunction with other techniques like NMR and Mass spectroscopy. The absorption of infrared radiation by a molecule causes changes in their vibrational and rotational energy levels, and hence IR-spectroscopy is also known as vibrational-rotational spectroscopy. IR spectroscopy provides a spectrum with a large number of absorption bands and hence provide plenty of information about the structure of a compound. Different bonds in the spectra correspond to various functional groups and bonds present in the molecule.

IR spectroscopy is the study of changes in vibration of the molecule when there is absorption of IR radiation by the molecule. Certain functional groups in the molecule give rise to bands at the same wavelength range irrespective of the structure of the molecule. This characteristic band gives useful structural information about the compounds when compared with the generalised charts of characteristic group absorbance bands. IR spectrum is characteristic of the molecule.

After absorbing IR radiation, the molecules of the compounds vibrate at many rates of vibration, giving rise to closely appearing bands in the IR spectrum, which is the fingerprint for the identification of the molecule. The characteristic functional groups and bonds of the compounds give rise to these bands, making it the fingerprint of the molecule.

THEORY

The IR absorption by the molecules causes excitation of both vibrational and rotational energies, and the spectrum appears as bands. The IR spectrum is otherwise called **vibrational spectroscopy**. Each vibrational energy change is accompanied by a number of rotational energy changes. These vibrational-rotational energy bands occurring between $4000 - 400 \text{cm}^{-1}$ are studied in IR spectroscopy.

Range of IR Radiation

IR range is divided into three regions.

$$\text{Near IR } (0.8 - 2.5 \mu m) \quad : \quad 13000 - 4000 \text{cm}^{-1}$$
$$\text{Far IR } (25 - 1000 \mu m) \quad : \quad 400\text{-}10 \text{ cm}^{-1}$$
$$\text{Mid IR } (2.5\text{-} 25 \text{ } \mu m) \quad : \quad 400 \text{cm}^{-1} \text{ to } 4000 \text{cm}^{-1}$$

In pharmaceutical analysis, the mid-IR region is used, which gives useful information about the vibration of molecules and the functional group present.

Bands in IR spectra are presented in wave numbers, and its unit is cm^{-1}. Wave numbers are reciprocal of the wavelength.

$$\text{cm}^{-1} = 10^4 / \text{ wavelength in } \mu m$$

The band intensity expressed in the wave number unit is easy to handle than the wavelength, either as absorbance or transmittance, because wave numbers are larger values, easy to handle and wavelengths show only small differences between functional group bands.

The mid-infrared spectrum extends from 4000 to 400 cm^{-1} and results from vibrational and rotational transitions. This region is most useful for the organic chemist since most organic molecules absorb in this region.

The mid-infrared can be divided into two regions, viz

- Functional group region (4000-1300cm^{-1})
- Fingerprint region (1300-600cm^{-1}).

Functional group region (4000-1300cm^{-1}): Most of the functional groups present in organic molecules exhibit absorption bands in the region **4000-1300 cm^{-1}**; hence this is known as the functional group region. In this region, each band can be assigned to a particular **deformation (bending)** of the molecule, the movement of a group of atoms, or the bending or **stretching** of a particular bond.

Fingerprint region (1300-600cm^{-1}): The region from **1300 cm^{-1} to 600 cm^{-1}** usually contains a very complicated series of absorptions. These are mainly due to molecular vibrations, usually bending motions that are characteristic of the entire molecule or large fragments of the molecule. Except for enantiomers, any two different compounds cannot have precisely the same absorption pattern in this region. Thus, **absorption patterns** in this region are unique for any particular compound; that is why this is known as the fingerprint region.

It is very difficult to assign individual bands in this region.

A typical IR spectrum is shown in figure 3.1.

Two molecules having the same functional group may show similar spectra in the functional group region, but their spectra differ in the fingerprint region. Therefore, both regions are very useful for confirming the identity of a chemical substance. This is generally accomplished by comparing the spectrum of an authentic sample. When two compounds show a good match between the IR spectra in all frequency ranges, mainly in the fingerprint region, it strongly indicates that they have the same molecular structures.

To understand the importance of fingerprint region to identify a compound, the example of propan-1-ol and propan-2-ol is given in Fig. 3.2 and 3.3.

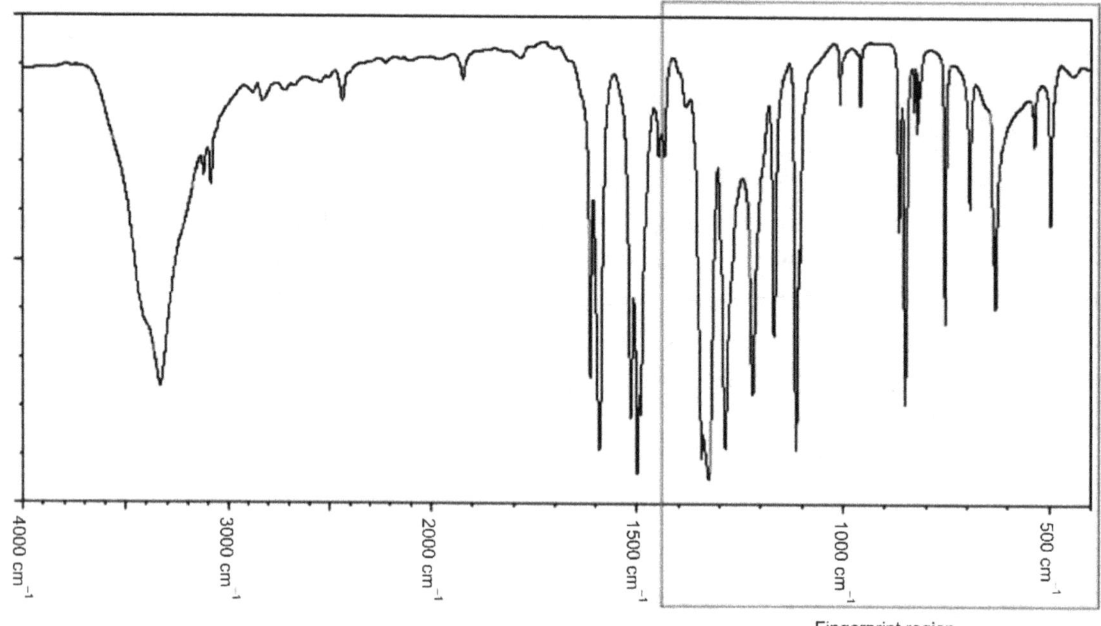

Fig. 3.1 A typical IR Spectrum.

Both the compounds are alcohols and contain precisely the same bonds. Now, if you compare the infra-red spectra of these compounds, the functional group region is very similar for both the compounds, but the fingerprint region is different like a fingerprint. Therefore, the fingerprint region can play a crucial role in identifying the compound.

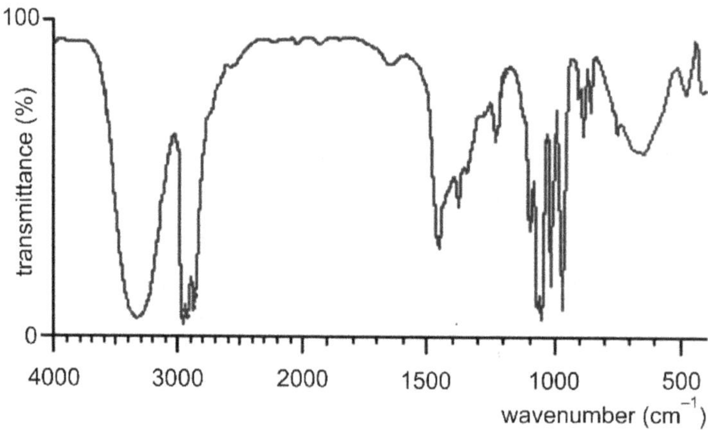

Fig. 3.2 IR spectra of Propane-1-ol – $CH_3CH_2CH_2OH$.

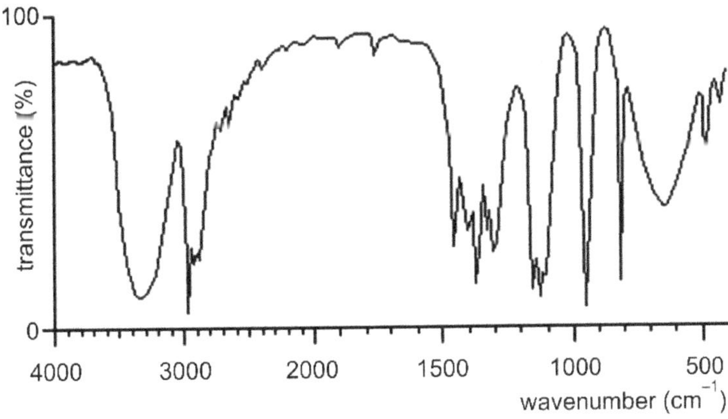

Fig. 3.3 IR spectra of Propane-2-ol CH_3CHCH_3
                                              |
                                             OH

CRITERIA FOR THE COMPOUND TO ABSORB IN IR REGION

(i) The functional groups in a molecule absorb IR radiation only when the **applied IR frequency** is **equal** to the **natural frequency** of vibration of the groups in the molecule

(ii) The IR radiation absorbed should cause a change in the dipole moment of the molecule

PRINCIPLE

The atoms or groups of atoms in a molecule are connected by bonds. These bonds are not rigid in nature, and the structure of the molecule acts like balls connected by a spring. Balls represent atoms, and the springs represent bonds. There are some vibrations of some frequency in the molecule, which is characteristic of the functional groups or bonds present in the molecule. These vibrational frequencies are called the natural frequency of vibration for that molecules. The applied IR radiation is absorbed when

Applied IR frequency = natural frequency of vibration of the molecule

As soon as the radiation is absorbed, a band appears in the spectrum.

The different functional groups or bonds in the molecules absorb different frequencies of IR radiation. Thus, for every functional group or bond, a characteristic band is observed.

Hence IR spectrum becomes the fingerprint of a molecule.

There are two types of molecular vibrations: stretching and bending.

Stretching is the change in bond length.

Bending is the change in bond angle.

Hooke's law can be used in the assignments of stretching frequencies of groups in molecules. The vibrating bond is compared to a physical model of a vibrating spring system. In application of Hooke's law, a bond and the connected atoms are considered to be a spring with two masses attached (Simple harmonic oscillator). The Hooke's law equation

$$\overline{v} = 1/2\pi c \times [f/(M_xM_y)/(M_x + M_y)]^{1/2}$$

Where $\overline{v}$ = vibrational frequency in cm^{-1}, C = velocity of light, f = force constant of bond (dynes/cm) and M_x and M_y = masses of (g) atom x and atom y respectively.

The equation indicates how the frequency $\overline{v}$, of the absorption changes as the properties of the system change.

An example of a molecule absorbing IR radiation is shown in figure 3.4. E.g., Methanol CH_3OH

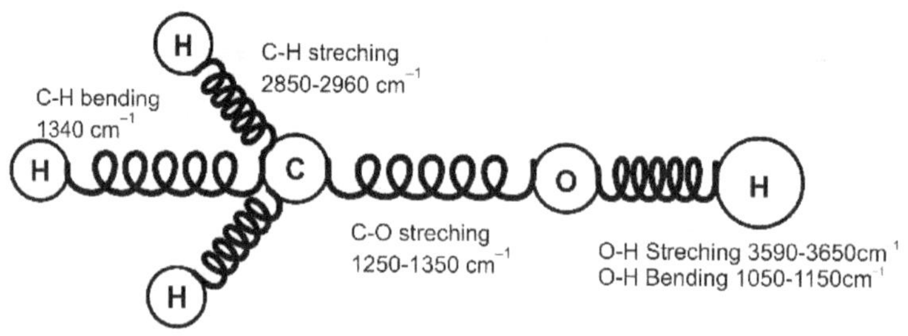

Fig. 3.4 Vibrations of a molecule after absorbing IR radiation.

Dipole Moment: The electrical dipole present in the molecule interacts with the electrical component of IR radiation. This interaction causes the change in the dipole moment of the molecule. The symmetrical diatomic molecules like O_2 and N_2 do not possess electric dipole, and they do not absorb IR radiation.

FUNDAMENTAL MODES OF VIBRATIONS IN POLYATOMIC MOLECULES

Molecular vibrations can be broadly classified into fundamental vibrations and non-fundamental vibrations. Under non-fundamental vibrations, we have overtones, combinations and Fermi resonance. The fundamental vibrations are further divided into **stretching and bending** vibrations. The **stretching vibrations** are further classified as **symmetric and asymmetric** stretching vibrations. The **bending vibrations** are classified as **in-plane** and **out of plane bending**, which are further divided into scissoring, rocking, wagging and twisting.

In fundamental vibrations, there is no change in the centre of gravity of the molecule. The principal fundamental vibrations are stretching and bending vibrations.

1. **Stretching vibration:** In these vibrations, the atoms move along the bond axis, and there will be changes in the **bond length** (increased/decreased) at regular intervals. These are two types of stretching vibrations

    - Symmetrical
    - Asymmetrical

    Symmetrical stretching: In this stretching, there is symmetrical increase or decrease of bond length occurs.

    Asymmetrical stretching: In this stretching, one bond length increases and the other bond length decreases.

2. **Bending vibrations:** A change in **bond angle** occurs in these vibrations. Bending vibrations are otherwise called **deformation vibrations** which are further classified as **in-plane bending and out of plane bending.**

    In-plane bending: Change in **bond angle** (bending) occurs in the **same plane**. Two types of in-plane bending are possible.

    (i)   **Scissoring:** In this type of bending, bond angle decreases

    (ii) **Rocking:** Both bonds are moving, but the bond angle is maintained.

    Out of plane bending: Bending of the bonds (bond angle change) occurs outside the plane of the molecule. The types are

    (i)   **Wagging:** Both the atoms move to the same side of the plane

    (ii) **Twisting:** In this type of bending, one atom moves above the plane and the other atom moves below the plane.

    The different types of vibrations are shown diagrammatically in Fig. 3.5.

NUMBER OF VIBRATIONS IN A MOLECULE

A compound's IR spectrum may show more than one vibrational absorption band. The number of these bands corresponds to the number of fundamental vibrations in the molecule, which can be calculated from the molecule's degree of freedom (DOF). Each atom in the molecule has three degrees of freedom (translational, rotational and vibrational). To describe the position of an atom in relation to other atoms in the molecule, the three Cartesian coordinates (x,y,z) are necessary. These three coordinates correspond to three degrees of freedom of the atom. Thus, there are 3n degrees of freedom for a molecule containing n atoms.

To find the number of vibrational degrees of freedom, the number of rotational and translational degrees of freedom should be deducted from total number of degrees of freedom.

For nonlinear molecules like H_2O

Total number of degrees of freedom = 3n = 3× 3 = 9 (n represents number of atoms)

3 degrees of freedom ---→ rotational changes

3 degrees of freedom -→ translational changes

So vibrational degrees of freedom = 3n – (3 + 3) = 3n – 6 = 9 – 6 = 3

Thus 3 vibrational changes are possible in a non-linear molecule.

For a linear molecule

For a linear molecule like CO_2

Total number of degrees of freedom = 3n = 3 × 3 = 9 (n represents number of atoms)

For rotational changes → 2 degrees of freedom

For translational changes → 3 degrees of freedom.

Therefore, vibrational degrees of freedom or fundamental vibrations = 3n – (3 + 2) = 9 – 5 = 4

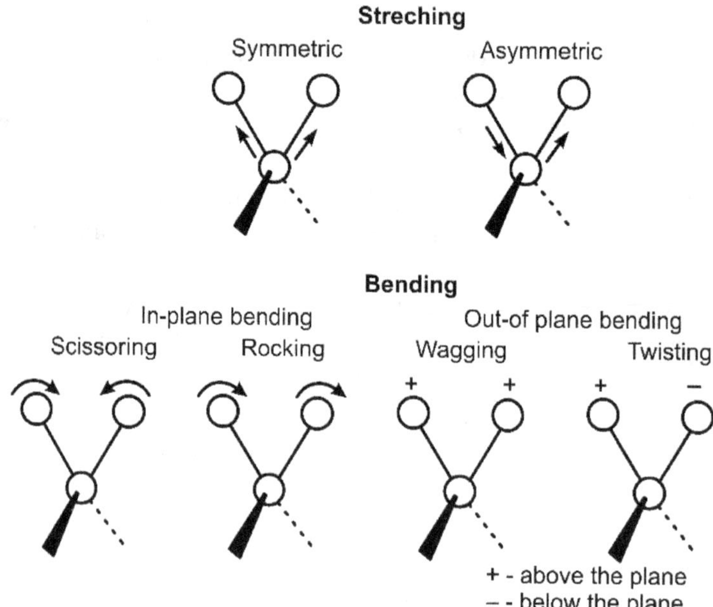

Fig. 3.5 Fundamental modes of vibrations of a molecule.

Thus four vibrational changes are possible in a linear molecule.

In the actual IR spectrum, the number of theoretical fundamental bands is rarely observed because some of the factors like overlapping of the spectrum, low resolution and weak peaks **may increase or decrease the number of bands**

Example CO_2 is linear molecule so expected to have 4 fundamental vibrations.

But actual spectrum shows only 2 vibrations ($666\,cm^{-1}$ and $2350\,cm^{-1}$).

Reasons

1. Symmetrical stretching of CO_2 is inactive in IR (no change in dipole moment)
2. Two scissoring/ bending vibrations are equivalent, and they appear at ($666\,cm^{-1}$) same frequency (degenerate)
3. Asymmetrical stretch appears as strong band at $2350\ cm^{-1}$

Thus, only two bands appear for CO_2 in the IR spectrum.

FACTORS INFLUENCING VIBRATIONAL FREQUENCIES

The value of vibrational frequency of a functional group calculated by Hooke's Law is not always equal to their observed value. The vibrational frequency of the group gets changed with the electronic and steric effects caused by other groups present in the surroundings. Following are some important factors affecting the vibrational frequency of a bond.

1. Bond order

2. Electronic effects

3. Resonance

4. Conjugation

5. Hydrogen bonding

6. Coupled vibrations and Fermi resonance

    1. Effect of Bond Order: Bond order affects the position of absorption bands.
The higher the bond order **larger** is the band frequency.
The frequency decreases with increasing atomic weight.
Frequency increases with increasing bond energy.

Table 3.1 Effect of bond energy and atomic masses on stretching frequencies.

Bond	Bond Energy [kcal (kJ)]	Stretching Frequency (cm^{-1})
Frequency dependence on atomic masses		
$C-H$ heavier	100 (420)	3000
$C-D$	100 (420)	2100 $\bar{\nu}$ decreases
$C-C$ atoms	83 (350)	1200
Frequency dependence on bond energies		
$C-C$	83 (350)	1200
$C=C$ stronger	146 (611)	1660 $\bar{\nu}$ increases
$C\equiv C$ bond	200 (840)	2200

    2. Electronic Effects: The frequency shift of the functional groups occurs due to electronic effects like **inductive effect, mesomeric effect and field effects.** Changes in the force constant or the bond strength are caused by these electronic effects leading to frequency shifts in absorption.

- **The +I effect of the alkyl groups** causes lengthening or weakening of the bond, and the absorption shifts to a **lower wave number.**
  E.g., Absorption wave number of Carbonyl group (C=O)
  (i) Formaldehyde, HCHO 1750 cm^{-1}
  (ii) Acetaldehyde, CH$_3$CHO 1745 cm^{-1}
  (iii) Acetone, CH$_3$COCH$_3$ 1715 cm^{-1}
- Substitution of electronegative atoms or groups with **–1 effect** shifts the absorption frequency to a **higher wave number.**
  (i) Acetone, CH$_3$COCH$_3$ 1715 cm^{-1}
  (ii) Chloro acetone, ClCH$_2$COCH$_2$ 1725 cm^{-1}
  (iii) Dichloroacetone, Cl$_2$CHCOCH$_3$ 1740 cm^{-1}
- Mesomeric effect also works along with inductive effect.

- The mesomeric effect involves the electrons in the **pie and nonbonding orbitals,** and it operates in general opposite to the inductive effect. These effects cannot be isolated from one another, and only one of them can be applied to determine approximate results.

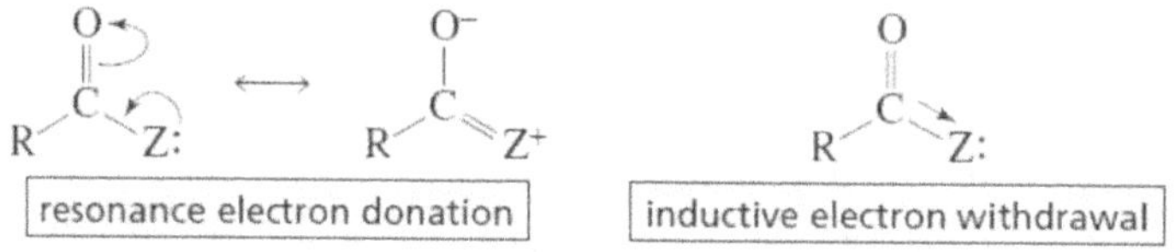

resonance electron donation

inductive electron withdrawal

3. Effect of Resonance

- Resonance also plays an essential role in frequency shifting.
- The C-O stretching frequency appears in the range of 1250-1050 cm^{-1}.
- If the C-O bond is in alcohol or ether, the stretching frequency will appear towards the lower end of the range.
- If this C-O bond is a part of a carboxylic group, the stretching frequency will occur at the higher end of the range.
- The position of the absorption varies because the bond in alcohol is a pure single bond, whereas the bond in a carboxylic acid has the partial double bond character that is due to resonance electron donation by the OH group in acids.
- Esters show stretches at both ends of the range because esters have two single bonds, one that is a pure single bond and the other bond has partial double-bond character.

CH$_3$CH$_2$—OH   CH$_3$CH$_2$—O—CH$_2$CH$_3$

~1050 cm^{-1} ~1050 cm^{-1} ~1250 cm^{-1} ~1250 cm^{-1} and ~1050 cm^{-1}

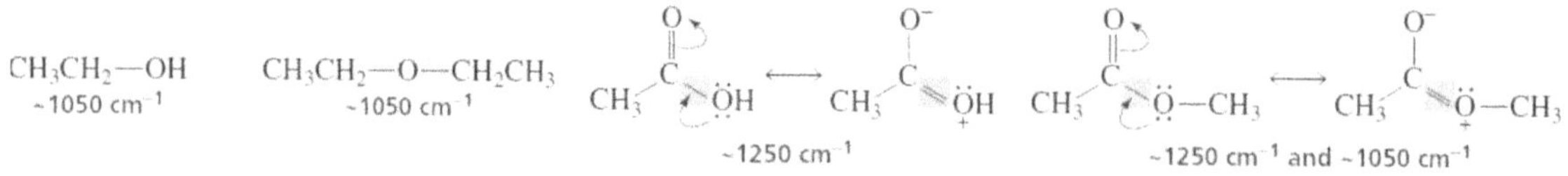

Fig. 3.6 Effect of resonance on Stretching.

4. Degree of conjugation

Conjugation shifts the absorption frequency to lower wave number

The presence of a greater degree of conjugation shifts the carbonyl absorption frequency to a relatively lower range than esters.

E.g., Benzamide and Phenylacetate.

The nitrogen atom is less electronegative than the oxygen atom in amide; It is more labile and participates to a greater extent in conjugation. So, Benzamide absorbs at a lower frequency than an ester.

5. Hydrogen Bonding

The presence of hydrogen bonding changes the position and shape of an infrared absorption band in both stretching as well as bending vibrations

The X-H stretching bands move to a lower frequency, usually increasing intensity and band widening.

The X-H bending vibration usually shifts to higher frequencies.

The stronger the hydrogen bonding, the greater is the absorption shift from the normal values.

The use of infrared spectroscopy can differentiate the two types of hydrogen bonding (Intramolecular and intermolecular).

The more concentrated the solution, the more likely it is for the OH-containing molecules to form intermolecular hydrogen bonds.

It is easier to stretch an O-H bond if it is hydrogen-bonded because the hydrogen is attracted to the oxygen of the neighbouring molecule.

Therefore, the O-H stretching of a concentrated (hydrogen-bonded) solution of alcohol occurs at about 3550 cm^{-1}, whereas the O-H stretching band of a dilute solution (with little or no hydrogen bonding) appears at 3650 cm^{-1} (Fig 3.7)

Hydrogen-bonded OH groups also have broader absorption bands, whereas the absorption bands of non-hydrogen–bonded OH groups are sharper.

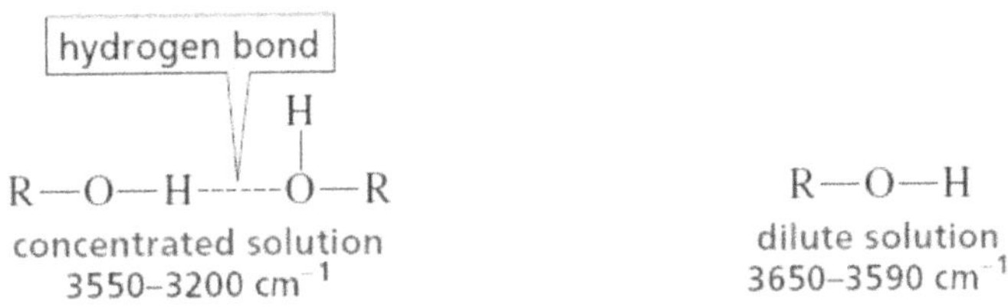

Fig. 3.7 Effect of hydrogen bonding on Stretching.

6. Coupled vibrations and Fermi resonance

For an isolated C-H bond, there is only one absorption frequency. But two absorptions occur for the CH_2 group corresponding to symmetric and asymmetric stretching vibrations. Asymmetric vibrations occur at a higher wave number when compared to symmetric vibrations. These vibrations are coupled vibrations because they occur at a different frequency than the isolated C-H vibrations. Because of these coupled vibrations, CH_3 also has stretching vibrations in different frequencies. Fermi resonance is a phenomenon that accounts for the shifting of the energies and intensities of absorption bands in infrared spectroscopy. Sometimes, it happens that two different vibrational levels have nearly the same energy. If such vibrations belong to the same species of molecules, then a mutual perturbation of energy may occur, resulting in the shift towards a lower frequency and the other towards a higher frequency. It is also accompanied by a substantial increase in the intensity of the respective bands. Fermi resonance was first discovered in carbon dioxide by Fermi, and it is also found in the vibrational spectra of aldehydes, where the fundamental stretching vibration of the C-H bond in the CHO group interacts with the second harmonic level(overtone) of the deformation vibration of the CHO group (2 × 1400 cm^{-1}). The result is a Fermi doublet with branches around 2830 cm^{-1} and 2730 cm^{-1}. It is essential for Fermi resonance that the vibrations connected with the two interacting levels be localized in the same part of the molecule.

INSTRUMENTATION

Common optical materials like glass or quartz absorb in the IR region. So, these materials are not used in the construction of an IR Spectrophotometer

The components of the IR spectrophotometer are

1. Radiation sources
2. Monochromators
3. Sample cells
4. Detectors

Table 3.2 Components of IR spectrophotometer.

Radiation source	Monochromators	Sample	Detectors
Nernst Glower	Alkali halides	Solid	Thermocouple
Incandescent lamp	Prisms	Liquid	Bolometer
Mercury arc	Gratings	Gas	Thermistor
Tungsten lamp	Filters		Golay cell
Globar source			Pyroelectric
Nichrome wire			Photoconductor

1. **Radiation sources**
   (a) **Incandescent lamps**
      1. It is made up of nichrome wire.
      2. They are used in near IR instruments.
      3. Has a low spectral emissivity.
      4. It is heated up to 1100K.
      5. Used to detect complex organic molecules
      6. It fails in far IR because it is enclosed in glass.
   (b) **Nernst Glower**
      1. It consists of a hollow rod about 2mm in diameter and 30 mm in length.
      2. They are made up of rare earth oxides of Zirconia, Thoria, Yttria etc.
      3. At room temperature, it is a non-conductor, and it is heated by external ways to bring it to conducting state.
      4. Heated up to 1000 to 1800°C
      5. Maximum radiation is provided at 7100 cm^{-1}

Advantages:
   1. IR radiation emitted over a wide wavelength range
   2. Steady emission of radiation intensity.
   3. Constant over a longer period of time.

Disadvantages:
   1. Frequent mechanical failure.
   2. Energy emitted in visible and near IR region

(c) Globar source

1. It is made up of sintered silicon carbide rod of 50mm in length and 4mm in diameter.
2. On heating up to a temperature between 1300 and 1700°C, it emits radiation in the IR region.
3. Maximum radiation is provided at 5200 cm^{-1}

Advantages:

1. It is self-starting
2. It can be controlled with a variable transformer as it's the temperature coefficient is positive.
3. More satisfactory for wavelengths longer than 650cm^{-1}

Disadvantages:

1. Less intense

(d) Mercury Arc

1. These lamps are used in the wave number region, less than 200 cm^{-1}(far IR region).
2. High pressures are used
3. In quartz mercury lamps, the quartz envelope emits radiation at shorter wavelengths, and the longer wavelengths are emitted by mercury plasma.

2. Monochromators

1. They convert polychromatic light into monochromatic light.
2. They must be constructed of materials which transmit the IR.

They are of 3 types.

(a) Metal halide prisms
(b) Sodium Chloride (NaCl) prisms
(c) Gratings

(a) Metal halide prisms

1. Prisms that are made up of Potassium bromide (KBr) are used in the wavelength region from 12-25μm.
2. Lithium Fluoride (LiF) prisms are used in the wavelength region from 0.2-6μm.
3. Caesium bromide (CeBr) prisms used in the wavelength region from 15-38μm.
4. Caesium bromide and Thallium Bromoiodide prisms are also used.

(b) Sodium chloride (NaCl) prisms

1. They are used in the whole wavelength region from 4000-650 cm^{-1}.
2. They have to be protected above 20°C because of their hygroscopic nature.

(c) Gratings

1. They offer better resolution at a low frequency than prisms.
2. Diffraction gratings of alkyl halides undergo etching in atmospheric moisture.

3. Sample cells

1. Sample cells made up of alkali halides like NaCl or KBr are used.
2. Aqueous solvents cannot be used as they dissolve alkali halides.
3. Only organic solvents like chloroform are used.
4. IR spectroscopy has been used to characterize solid, liquid, gas samples.

4. Detectors: The various detectors used in IR spectrophotometers are Bolometers, thermocouple, Thermistors, Golay cell and pyroelectric detectors

Bolometer

1. The principle of operation is that the electrical resistance of a metal increases with an increase in temperature.
2. It consists of a thin metal wire
3. When radiation falls on this conductor, its temperature changes.
4. The change in resistance measures the radiation falling on the bolometer.
5. Bolometer is used as one arm of Wheatstone bridge with a similar metal strip in balancing arm.
6. The bridge remains balanced till no radiation falls on the bolometer
7. When IR radiation falls on the bolometer, the bridge is unbalanced, with a change in resistance leading to current flow indicated by the galvanometer.
8. The intensity of radiation falling on the detector is measured by the current flowing through the galvanometer.
9. The bolometer response time is 4 msec

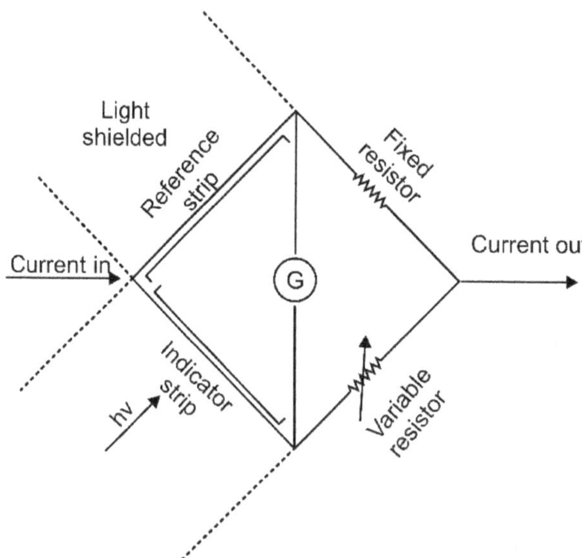

Fig. 3.8 Schematic diagram of the bolometer-Wheatstone bridge.

Thermocouple

1. Two dissimilar metal wires are joined head to tail.
2. One end is exposed to IR radiation (hot junction), and the other junction is insulated(cold junction)
3. The hot junction which is exposed to IR radiation increases in temperature, and the cold junction is maintained at a constant temperature
4. The temperature difference between the two junctions creates a potential difference that is directly proportional to the radiation falling on the hot junction
5. The response time is about 60msec

This detector is based on the principle that the difference in temperature between the junctions is directly proportional to the current flowing through the system, which in turn depends on the intensity of radiation.

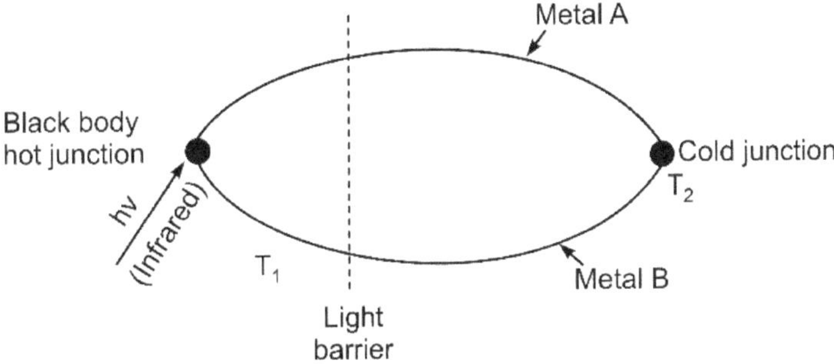

Fig. 3.9 Schematic diagram of a thermocouple.

Thermistors

1. These detectors are made up of sintered oxides of Mn, Co and Ni.
2. The principle here is as the temperature of the mixture increases, the resistance decreases
3. Response time is 4 seconds.

Golay cell

1. Used in commercial spectrophotometers
2. It consists of a metal cylinder with a blackened metal plate at one end and a flexible metallised diaphragm at the other end.
3. The cylinder is filled with Xenon gas.
4. IR radiation falling on the metal plate heats up the gas and expands it. The resulting pressure increase deforms the diaphragm.
5. When light falls on the diaphragm from the source, it reflects the light to a photocell.
6. The motion of the diaphragm results in changes in the output of the cell, which is recorded as a signal.
7. Response time 0.01 sec

Advantages:

1. Wide wavelength range
2. Linear response

Disadvantages:
1. Bulky
2. Expensive

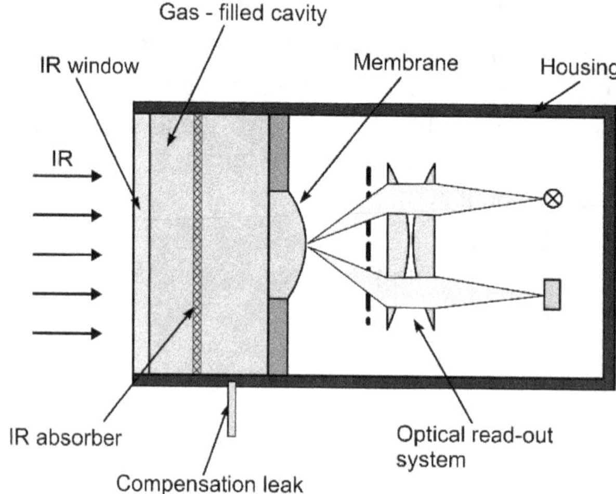

Fig. 3.10 Golay cell.

Photoconductivity cell
1. It is a non-thermal detector.
2. It comprises a thin layer of lead sulphide or lead telluride deposited on glass sealed in an evacuated glass envelope.
3. The conductance of semiconductors increases when IR radiation falls on it, leading to the flow of currents.
4. Its response time – 0.5 msec

Advantages:
1. Greater sensitivity
2. Faster response.

Disadvantages
1. Range limited to near IR radiation at room temperature.

Pyroelectric Detectors

A dielectric material, when placed in an electrostatic field, is polarized and this polarization depends on the dielectric constant of the material. If the polarization persists even after the removal of the electric field, then the material is said to be pyroelectric. This residual polarization is temperature-dependent.

1. These detectors are made by sandwiching the pyroelectric material between two electrodes to give a temperature-dependent capacitor.
2. The incident IR radiation causes temperature changes leading to a change in capacitance of the material, which can be measured.
3. Commonly used pyroelectric material is Triglycerine Sulphate (TGS)
4. Deuterated TGS can be used at room temperature, and it is used in FTIR.

Advantages:
1. Fast response
2. Multiple scanning
3. Used in FTIR

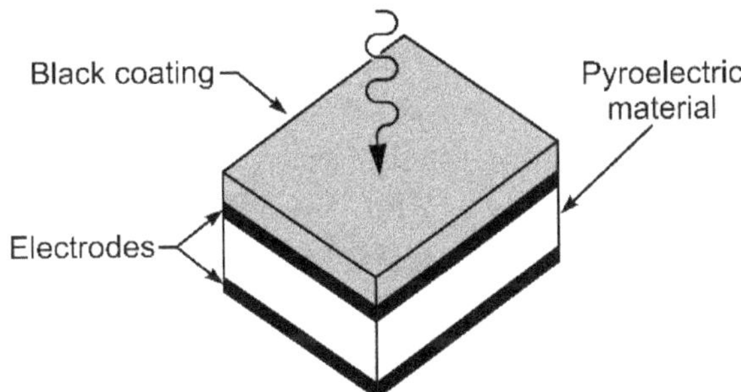

Fig. 3.11 Pyroelectric detectors.

Semiconductor detectors: Semiconductors are insulators, and they become conductors when exposed to radiation, which causes a change in electrical resistance.

Materials like telluride, indium antimonide and copper or mercury doped with germanium are used as semiconductor detectors.

5. **Recorder:** A recorder is used for recording the spectrum.

TYPES OF INSTRUMENTS

SINGLE BEAM IR SPECTROPHOTOMETER

In a single beam instrument, the radiation source passes through the sample and then through the fixed prism and Littrow mirror to select the desired wavelength. Finally, the radiation falls on the detector.

Thus, in IR instrumentation, the intensity of radiation is measured after it passes through the sample and monochromator.

Sample cells are present before monochromators to diminish stray/scattered light problems.

Disadvantages:
1. The intensity of emitted radiations is not uniform.
2. Solvent bands appear in the spectrum.

DOUBLE BEAM SPECTROPHOTOMETER

The light energy is split into a reference beam and sample beam. They pass through the respective cells and reach the detector through the monochromator. When there is no sample in the sample cell, there is no absorption, and a steady signal reaches the detector. When there is absorption, there is a decrease in the intensity of light reaching the detector producing a signal.

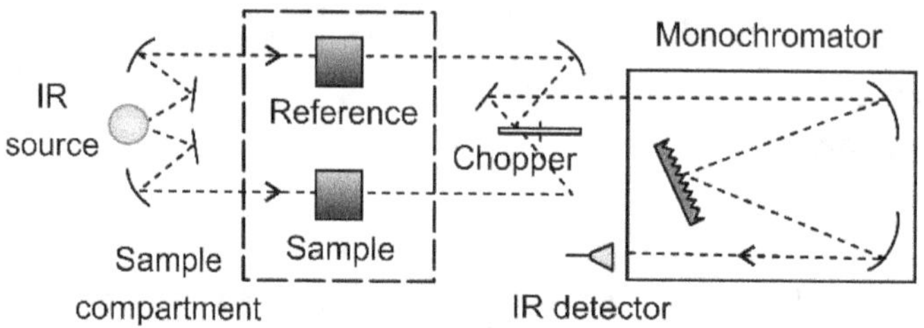

Fig. 3.12 Schematic diagram of Dispersive IR spectrophotometer.

FOURIER TRANSFORM IR SPECTROPHOTOMETER

This instrument depends upon the interferences of various frequencies of radiation to produce a spectrum.

FTIR spectrophotometer consists of an interferometer which consists of four optical arms at right angles to each other. It consists of a beam splitter at the centre. Radiation from the source passes through the first arm, split by the beam splitter into two perpendicular beams of equal intensity. These split beams enter into the other two arms of an interferometer. In these arms, one arm has a fixed mirror, and the other arm has a movable mirror. The motion of the moving mirror makes the total path length variable from the pathlength taken by the fixed mirror beam. These two beams meet again at the beam splitter, they recombine, and their difference in path length causes constructive and destructive interference to give the interferogram.

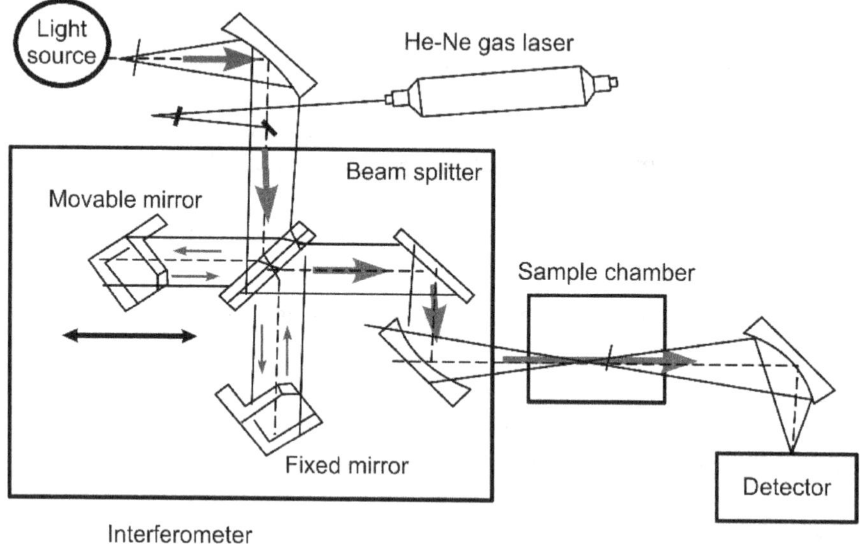

Fig. 3.13 Schematic diagram of FTIR spectrophotometer.

The recombined beam passes through the sample. The sample absorbs its characteristic wavelengths, and the detector detects the variation in the intensity of the transmitted light. The spectrum is recorded as intensity versus time for all wavelengths simultaneously.

A mathematical function called Fourier Transform is used to convert the intensity versus time spectrum into intensity versus frequency spectrum.

In practice, it is difficult to control the speed of the moving mirror and hence it is controlled by a laser beam. The laser beam serves as a reference for the operation of the instrument.

Table 3.3 Difference between FT IR and dispersive IR.

DISPERSIVE IR	FTIR
1. Many moving parts lead to wear and tear.	Only one mirror moves during the experiment.
2. Calibration against reference spectra is required to measure frequency.	The laser provides high-frequency accuracy.
3. Stray light causes errors in the readings.	Stray light does not affect the detector.
4. To improve resolution, only a small amount of IR radiation is allowed to pass through the slits	A larger beam is used. (internal calibration). Data collection is clear.
5. Only radiation of narrow frequency falls on the detector at a given time.	All frequencies of radiation fall on the detector simultaneously.
6. The scan speed is slow.	Rapid scan
7. Sample shows the thermal effect	No thermal effect.
8. Lesser accuracy.	More accuracy
9. Sensitivity is less.	Sensitivity is more.

SAMPLE HANDLING TECHNIQUES IN IR

There are no good solvents that are transparent throughout the IR region. Because of this, sample handling or preparation becomes a tedious and time-consuming part of IR analysis. Sample handling of different types of samples are given below

Sampling of solids

Four techniques are employed for preparing solid samples.

1. Solids run in solution
2. Solid films
3. Mull technique
4. Pressed Pellet technique
5. ATR crystal technique. (evanescent wave)
1. **Solids run in solutions:** If the sample can be dissolved in a suitable solvent, then the cells for liquids can be used. The drawback of this method is the limited number of solvents that do not absorb in the IR region. Examples of solvents: chloroform, carbon tetrachloride, acetone and cyclohexane.
   Disadvantages:
   1. There is no solvent available which do not absorb throughout the IR region.
   2. Interaction of solvent with the solute should be monitored (H-bonding compounds).
   3. The solvent should not absorb in the working IR range.

2. **Solid film:** The sample is dissolved in a non-aqueous solvent, non-reacting, IR transparent in the working IR range. A drop of this solution is placed on an alkali metal plate (e.g., NaCl/ KCl plate), and the solvent is evaporated to dryness, leaving a thin sample film. This technique is suitable for polymer resin and amorphous solids. For non-crystalline solids, a thin homogenous film is deposited on the plate, which can be directly mounted and scanned directly.

   Advantages: This technique is useful for qualitative analysis.

   Disadvantages: Not suitable for quantitative analysis.

3. **Mull technique:** A small quantity of sample (2-4mg) is thoroughly ground in a clean mortar until the powder is very fine. The mulling agent (mineral oil or Nujol) is introduced in small quantities, just sufficient to make a mixture approximating the consistency of toothpaste.

   Then the mixture is transferred to the mull plates & the plates are squeezed together to get the desired thickness of the sample between IR transmitting windows which is mounted in the path of the IR beam, and the spectrum is obtained.

   Disadvantages

   1. Although Nujol is transparent throughout the IR region, yet it has an absorption maxima at 2915, 1462, 1376 & 719 cm^{-1}.
   2. So, when the IR spectrum of a solid sample is taken in Nujol mull, absorption bands of a solid sample that happen to coincide with the absorption bands of the Nujol mull will be hidden (but others will be clearly seen in the IR spectrum)
   3. This method is good for qualitative analysis but not for quantitative analysis.
   4. If hydrocarbon bands interfere, Fluorolube, a halogenated polymer, can be used.

4. **Pressed pellet technique:** In this technique, a small amount of finely ground solid sample is intimately mixed with about 100 times the weight of powdered Potassium bromide in a vibrating ball mill. This finely ground mixture is then pressed under very high pressure (25000 p sig) in an evacuable die or Minipress to form a small pellet (about 1-2 mm thick and 1cm in diameter). (10,000 -15,000 pounds). The resulting pellet is transparent to IR radiation and is used as such to record the spectrum.

   The powder (KBr + sample) is introduced in between the two bolts, and the upper screw A is tightened until the powder is compressed to a thin disc. After compressing the sample bolts, A & A$'$ are removed, and a steel cylinder with a pellet inside it is placed in the path of the beam of IR spectrometer, and a blank KBr pellet of identical thickness is kept in the path of the reference beam.

   Advantages of this technique over mull technique

   1. The bands due to mulling agent in mull technique are not present in the KBR pellet technique
   2. KBr pellets can be stored for longer periods of time.
   3. It can be used for quantitative analysis, as the concentration of the sample can be adjusted in pellets
   4. The spectrum resolution in KBr is superior to that obtained with mulls.

Disadvantages

1. The –OH group of moisture present in the sample gives a band at 3450 cm^{-1}.

2. The high pressure used in pellet formation may bring about polymorphic changes in crystallinity in the samples (Especially inorganic complexes), which may cause complications in the IR spectrum. Even substitution of the ligand by bromide may be possible in inorganic complexes.

3. This method is not applicable for some polymers which are difficult to grind with KBr.

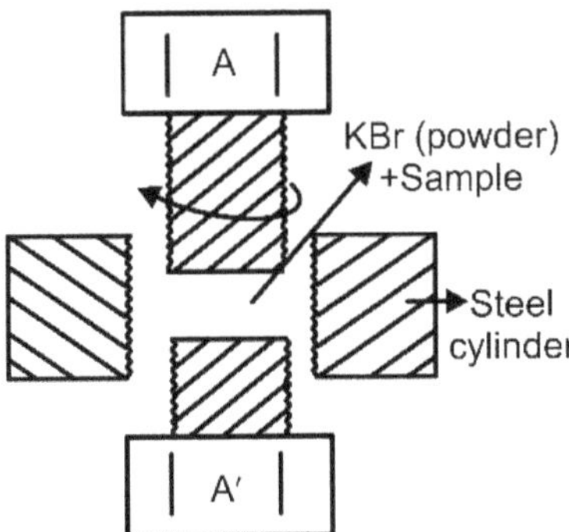

Fig. 3.14 Device used for potassium bromide discs.

5. **ATR (Attenuated Total Reflection) crystal technique:** Without further preparation, samples can be analysed directly in a solid or liquid state. The sample is in contact with the surface of an ATR crystal, which is made up of a material with a higher refractive index. E.g., silver chloride, zinc selenide, zinc sulphide, germanium silicon and diamond. Liquid samples are simply poured over the crystal, and solid samples are clamped onto it. When IR radiation passes through the sample, an evanescent wave is formed, and this absorption is detected by the detector.

Sampling of liquids: Liquid samples are taken in sample cells made up of NaCl, KBr or ThBr and the IR spectra is measured directly.

A thin layer of the sample with a thickness of about 0.01 - 0.05 mm is used.

For double beam instruments, are matched cells with the same thickness used for reference and sample.

The cells should be protected from moisture.

Sampling of gases: In the gas sample cell, the light path is made of KBr or NaCl.

They are larger cells with longer pathlength about 10cm to 1m long to compensate for the small number of molecules in the sample gas. Multiple reflection pathlengths are used to make the pathlengths as long as 40 cm as shown in figure 3.15.

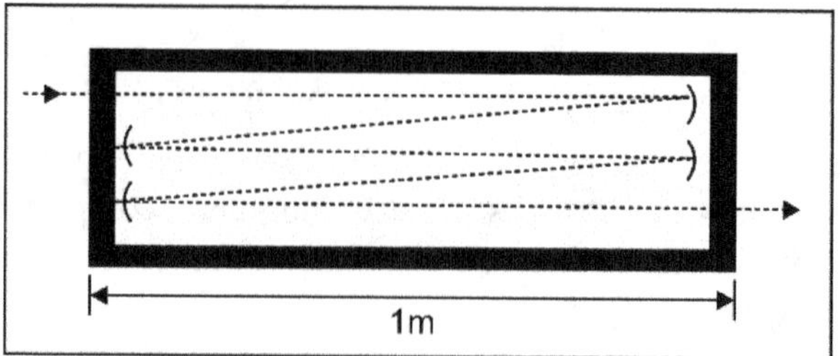

Fig. 3.15 Gas absorption cell.

The gas sample cell is a cylinder of glass or metal. Both ends of the cylinder are closed with IR transparent materials like NaCl/KBr.

They are equipped with valves or stopcocks for the introduction of a sample.

SOLVENTS

1. No single solvent is transparent throughout the mid-IR region.
2. Water and alcohol are rarely employed because they absorb strongly and attack the sample cells, which are made up of alkali metal halides.

Table 3.4 IR Absorption of Some Common Functional Groups.

Functional Groups	Range (cm^{-1})
C-H stretching (alkane)	2960-2850
C-H bending	1340
C-H stretching (alkene)	3040-3010
C-H stretching (aromatic)	3030
C-H bending	700-850
C=C stretching (alkene)	1680-1620
C≡C stretching (alkyne)	2100-2200
C=C stretching (aromatic)	1450-1600
C=O stretching (ketone)	1705-1725
C=O stretching (aldehyde)	1720-1740
C=O stretching (ester)	1735-1750
C=O stretching (acid)	1700-1725
C=O stretching (amide)	1650-1700
O-H stretching (Free)	3590-3650
O-H bending(alcohols)	1050-1150
C-O stretching	1250-1350

Table 3.4 *Contd...*

Functional Groups	Range (cm^{-1})
O-H bending(Phenols)	1200
C-O stretching	1310-1410
N-H stretching	3400-3500
N-H bending	1500-1650
C-N vibrations	1000-1400
C≡N stretching	2240-2260
C=N stretching	1630-1690

APPLICATIONS

1. **Identification of functional group and structural elucidation:** The peaks corresponding to every functional group in the sample appears in the group frequency region – 4000-1500 cm^{-1} which is useful in structural elucidation. The fingerprint region is useful in identifying the sample, including the isomer.

2. **Qualitative estimation:** The structure of the sample can be confirmed if the IR spectrum of the sample and standard are identical (spectral matching).

3. **Identification of impurities in a drug sample:** Impurities give extra peaks in the spectrum when compared with the spectrum of the pure drug thus presence of impurities can be confirmed.

4. **To study the progress of reaction:** The disappearance of a characteristic absorption frequency of a reactant shows the progress of the reaction. Spectra of samples taken at different time intervals can be used in studying the progress of the reaction.

5. To identify and find the ratio of geometrical isomers (cis, trans) in a mixture

6. Identification of tautomers.

7. To find the shape of symmetry of a molecule.

8. The presence of water in a sample can be identified using the IR spectrum, which shows characteristic absorption of water in 3 regions 3600-3200 cm^{-1}, 1650cm^{-1}, 600-300 cm^{-1}.

9. **Quantitative analysis:** The quantitative estimation of the substance can be determined in pure form or as a mixture of two or more components.

Two methods are used: 1. Cell in cell-out method. 2. Baseline method.

In the cell-in-cell-out method, the unknown concentration of sample is determined using calibration curve of the standard or using beers law by comparing log I_0/I_t (absorbance) of the standard and sample.

In the baseline method, tangents are drawn, I_0 and I_t are measured, absorbance is plotted against concentration, and unknown concentration is determined.

CHAPTER 4

Flame Photometry

INTRODUCTION

Flame photometry is a branch of spectroscopy where radiation emitted by the metallic atoms is measured. It is otherwise called **Flame atomic emission spectroscopy**. The flame photometry principle is based on the measurement of emitted light intensity when a metal is introduced into the flame. The colour of the emitted light is characteristic for each element and is used for qualitative analysis (identification). The intensity of the emitted radiation is directly proportional to the concentration of the metal and this forms the basis of quantitative estimation.

PRINCIPLE

When a sample solution containing metal salt is nebulised through a flame, a series of processes occur, which are given below

1. Spray leaves fine droplets of metal salt solution in flame – **nebulisation**

2. The solvent gets vaporised due to the thermal energy of the flame, leaving a fine residue of solid salt – **desolvation**

3. The salt is vaporized and gets converted to gaseous state – **vaporisation**

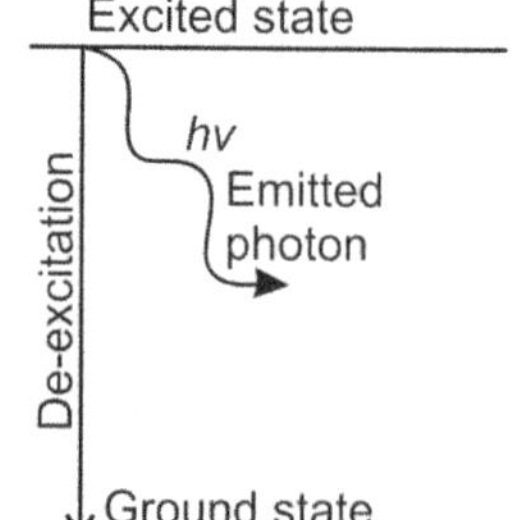

Fig. 4.1 Diagrammatic representation of flame photometry principle.

4. The gaseous molecules are dissociated to give neutral atoms, which are excited by the thermal energy of the flame – **atomization and excitation**

5. The excited atoms are unstable due to the high energy of the excited state, and they emit radiation while returning to the unexcited ground state - **Emission**.

The wavelength of emitted radiation is characteristic of the element emitting it. The wavelength of emitted radiation is used to identify the element (Qualitative analysis). As the intensity of the emitted radiation is directly proportional to the concentration of the element, the intensity is measured and used for quantitative analysis. The measurement of the emitted radiation forms the basis of flame photometry and the steps involved are given in the flow chart.

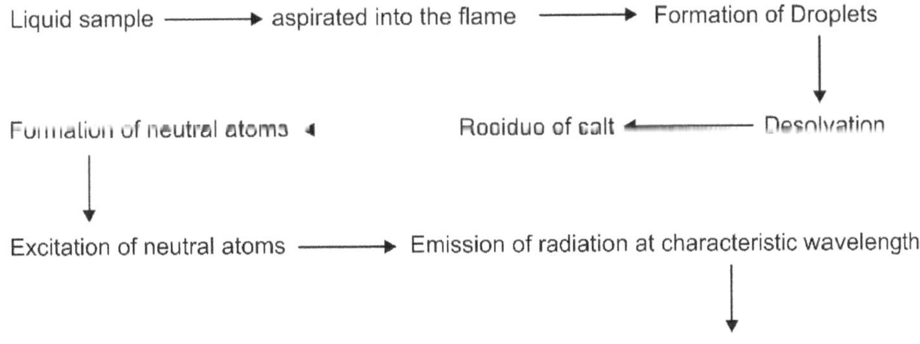

Boltzmann distribution governs the fraction of free atoms that are thermally excited and is given as

$$\frac{N^*}{N_0} = Ae^{-\Delta E/kT}$$

where

    N^* – No. of excited atoms

    N_0 – No. of remaining ground state atoms

    A – Constant for the given element

    ΔE – Difference in energy between the ground state and excited state.

    k – Boltzmann constant

    T – Flame temperature

Thus from Boltzmann distribution, it is understood that the energy of excitation is dependent on flame temperature. Thus, flame is useful only for the elements that require low excitation energy (like alkali metals and alkaline earth metals), i.e. Group I A (Li, Na, K) and Group II A (Ca,Mg etc.).

Since the emitted radiation is characteristic for each element, the wavelength of emitted radiation is different for different elements

    For E.g.,   Sodium emits yellow colour radiation at 589 nm

            Potassium emits lilac colour at 767 nm

            Calcium emits brick red colour at 422, 554 and 626 nm

            Lithium emits red colour radiation at 670 nm

The intensity of the emitted radiation is directly proportional to the concentration of the element present in the sample.

INSTRUMENTATION

The basic components of flame photometer are

1. Sample delivery systems
2. Source
3. Monochromators
4. Detector
5. Readout device

Table 4.1 Summary of Instrumentation.

Sample delivery systems	Source	Monochromators	Detector
(i) Nebulizer, (ii) Aerosol modifier, (iii) Flame or Atomiser	Burners 1. Mecker burner 2. Total consumption burner 3. Premix laminar flow burner 4. Lundergarph burner 5. Shielded burners	Filter wheel Quartz Prism Grating	Photovoltaic cells Phototubes Photomultiplier tube

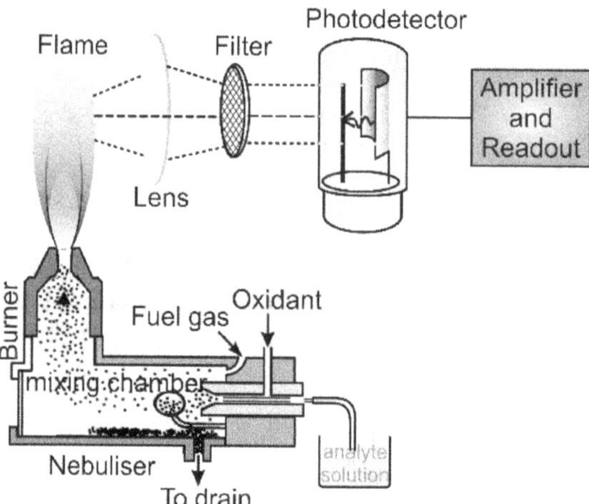

Fig. 4.2 Diagrammatic representation of Flame Photometer.

1. Sample Delivery Systems

Three components are there in the sample delivery system.

(i) Nebulizer: The conversion of sample solution into a mist of finely divided droplets (aerosol) is called nebulisation. The sample is carried to the atomizer from the nebuliser.

There are three types of classification of nebulisers. The classification is based on

1. The nebulisation process
   Pneumatic or ultrasonic nebulisers.
2. Feeding of the sample
   Suction, gravity-fed, controlled flow reflux nebulizer
3. Relative position of capillaries for the aspirated liquid and nebulising gas and the aspirated liquid
   Angular and concentric nebulisers

(ii) Aerosol modifier: The large droplets are removed from the stream, and this modifier allows only the droplets smaller than the defined size to pass through.

(iii) Flame or Atomiser: The atomizer performs the conversion of vaporized analyte into free atoms

The different types are
1. The type which introduces the sample directly into the flame.
2. The type which introduces the sample in the chamber of an atomiser.

STRUCTURE OF FLAME

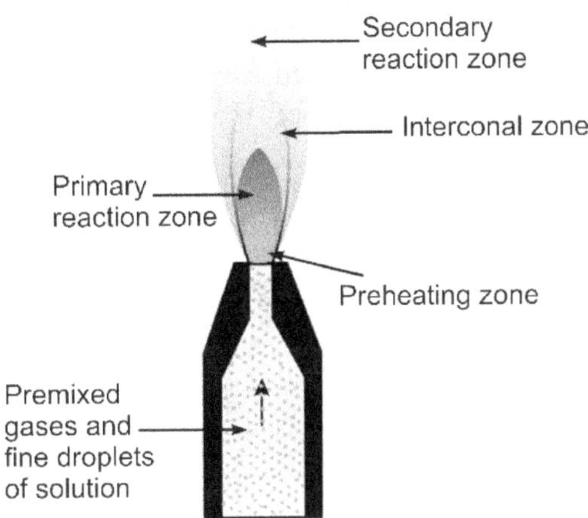

Fig. 4.3 Structure of Flame.

The flame is divided into
1. Preheating zone
2. Primary reaction zone/inner zone
3. Interconal zone
4. Secondary reaction zone

The maximum temperature is obtained above the tip of the primary reaction zone, and this zone is used for flame photometry.

Functions of flame
(i) To desolvate the liquid droplets from the sample forming residue.

(ii) To decompose the residue to form atoms

(iii) To excite the atoms so that they can emit radiation when returning to the ground state.

The temperature of the flame is responsible for all these events, and the factors controlling the temperature of the flame are

(i) Nature of fuel, oxidant and their ratio

(ii) Nature of the solvent used to dissolve the sample.

(iii) Amount of solvent entering the flame

(iv) Type of the burner

REQUIREMENTS FOR FLAME

1. Proper temperature should be maintained, so that the energy supplied is lesser than the ionization potential of the analyte atom
2. Temperature should be maintained constant throughout the procedure
3. No fluctuations in burning should be there.

Fuel and oxidant gases

The optimum temperature for the flame is important because the too low temperature may not be able to excite the atoms, and too high temperature may lead to ionization of the atoms causing the population of excited atoms to decrease (decreased intensity of emission). The different combinations of fuel and oxidant gases used are given in table 4.2.

Table 4.2 The different combination of fuel and oxidant gases.

Fuel	Temperature in oxygen	Temperature in air
Methane	2700	2000
Propane	2800	1925
Butane	2900	1010
Hydrogen	2780	2100
Acetylene	3050	2000
Cyanogen	4580	-----

Various hydrocarbon gases give satisfactory temperatures for metal analysis. Ordinary cooking gas can give a temperature of about 1700°C. Acetylene flame can give a temperature of about 2000-2300°C and is used to determine Li/Mg metals.

2. **Source-Burners:** The burners are used to spray the sample solution into the flame after mixing with fuel and oxidant. The different types of burners are

    1. Mecker burner

    2. Total consumption burner

    3. Premix laminar flow burner

    4. Lundergarph burner

    5. Shielded burners

1. **Mecker burner:** This burner uses natural gas and oxygen and is not used nowadays. It produces low temperature and a non-homogenous flame.

2. **Total consumption burner:** In these burners, hydrogen and oxygen gases are used as fuel and oxidant gases, respectively. The liquid sample is aspirated into the flame through a capillary tube. The oxygen and hydrogen enter through the side tubes and are burnt at the top of the burner to produce a flame. As soon as the sample enters the flame, it is vaporised, atomized and then excited.

   It is called a total consumption burner since the entire sample entering the capillary is consumed.

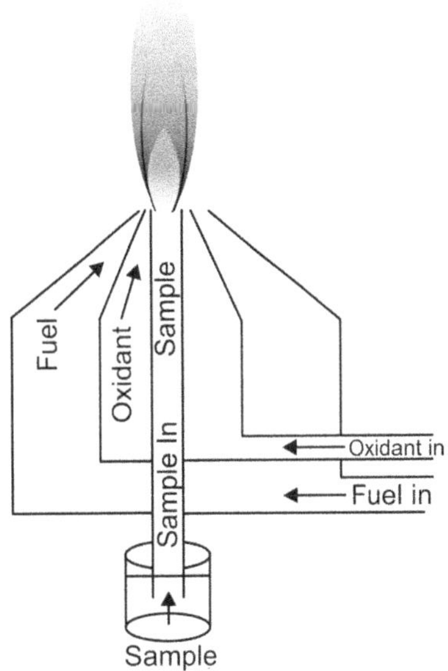

Fig. 4.4 Total consumption burner.

Advantages

1. Simple design and large sample size enhances sensitivity
2. Entire sample consumed
3. No explosive hazard.

Disadvantages

1. Large droplets not decomposed completely
2. Non- uniform and non-homogenous flame
3. Fluctuations in flame intensity
4. Viscosity of the sample affects the rate of sample introduction
5. Short path for the solute decreases sensitivity

3. **Premix laminar flow burners –most widely used burner:** In this burner, the sample, fuel and oxidant gas are mixed thoroughly and then introduced into the flame. The flow of the gases is laminar and not turbulent as in the total consumption burner. Only a small fraction of the sample droplets reach the flame. Easy decomposition of the sample takes place, leading to efficient atomisation. Large droplets are drained off the outlet.

Advantages

1. Noiseless, non-turbulent and stable flame.
2. Efficient atomisation
3. Clogging free
4. Long path provides sensitivity

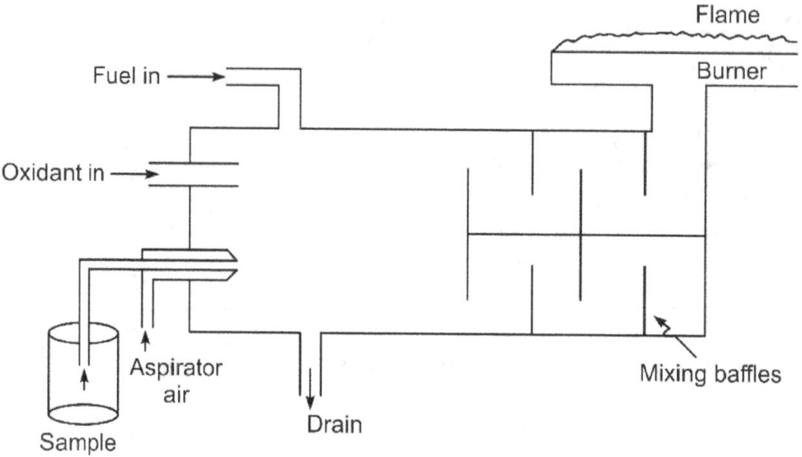

Fig. 4.5 Premix laminar flow burners.

Disadvantages

1. Only a small quantity of the sample reaches the flame
2. Rate of sample introduction is slow.
3. Explosive hazard during mixing

3. Filters and monochromators

Prism: A quartz prism is used as quartz is transparent over the entire region.

Grating: Grating monochromators can also be used

A detailed description of filters and monochromators are given in UV Visible spectroscopy.

A filter wheel present in a simple flame photometer contains several filters (Filters for each element) like Ca, Li, Na, or K. Specific filter will be selected based on the analyte to be determined.

4. **Detectors:** The emitted radiations are in the visible region. Hence detectors like **photovoltaic cells or phototubes** are used in the flame photometer. Flame spectrophotometer uses a photomultiplier tube as a detector. These detectors are explained in UV spectroscopy.

5. **Readout device:** The detector signal is displayed in a digital readout device. The emission at a specific wavelength will be displayed.

INSTRUMENTS

1. Flame Photometer

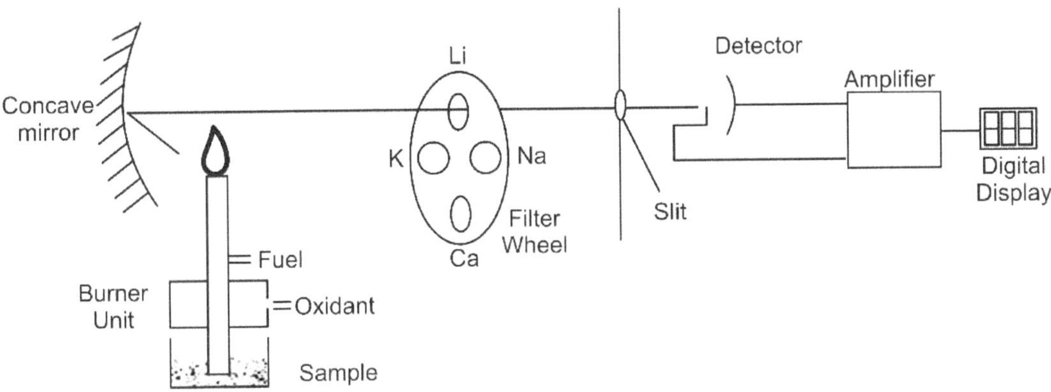

Fig. 4.6 Schematic diagram of a Flame photometer.

Working

The solution to be analysed is sprayed as fine droplets (aerosol) into the burner's flame after mixing with fuel and oxidant gases.

Air or oxygen and the fuel gas are compressed to a pressure of about 0.5kg/sq.cm using a compressor. This high pressure helps in the spraying of samples in fine droplets. This process is called nebulisation.

Next, the desolvation process occurs, followed by vaporization and atomisation.

The atoms formed are excited by the thermal energy of the flame.

These excited atoms are unstable, and they emit excess energy in the form of light to come back to the ground state.

A concave mirror focuses the light to the filter wheel containing Li,Na,K and Calcium filters. The specific filter for the analyte is selected, and the light is allowed to pass through it.

The emitted radiation is directly proportional to the concentration of the substance and is measured by the detector and displayed in the readout device.

2. **Flame Spectrophotometer:** The filter used in the flame photometer is replaced with a prism or grating monochromator in the flame spectrophotometer. A Photomultiplier tube is used as a detector.

3. **Double Beam Spectrophotometer:** This instrument is otherwise called as **internal standard flame photometer**. Lithium is used as an internal standard, and an equal concentration of lithium is added to both standard and sample. The emitted radiation is divided into two parts and measured by two different detectors. One part of the beam is due to lithium (internal standard), and the other is due to the sample. The ratio of the emission intensity of the analyte to the internal standard is recorded.

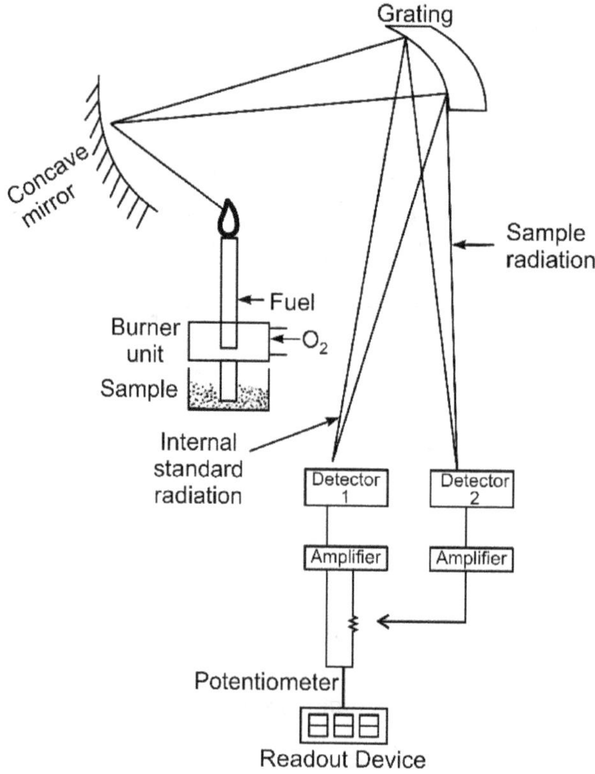

Fig. 4.7 Schematic diagram of a Double beam Flame spectrophotometer.

Advantages

1. Use of internal standards cancels out the error due to the fluctuations in flame.
2. Errors due to differences in viscosity and surface tension are also minimised.

APPLICATIONS

1. **Qualitative analysis:** Flame photometry is used to detect group I and group II elements of the periodic table. The sample is identified by the technique called peak matching. In this technique, a minimum of 3 peaks of emission spectrum should match the standard.

   E.g.: calcium emits radiation at 422,544, 626 nm. If the sample spectrum shows matching peaks in these wavelengths, then the sample is confirmed as calcium.

2. **Quantitative analysis:** Flame photometry is used for rapid quantitative estimation of Group I and II elements like lithium, calcium, sodium and potassium in samples.

   Some of the quantitative applications are determination of

   1. Concentration of calcium in serum
   2. Concentration of calcium, sodium, potassium in the urine

3. Concentration of calcium, sodium, potassium and magnesium in IV fluids and oral rehydration salts.

4. Estimation of potassium chloride in syrup.

5. Serum lithium concentration in Therapeutic drug monitoring

Experimental procedure

There are four methods available for quantitative estimation

1. Direct comparison method

2. Calibration curve method

3. Standard addition method

4. Internal standard method

1. **Direct comparison method:** The emission of the sample solution is compared with standard solution emission.

   Advantages:

   1. Speed of analysis

   2. No need to prepare more standards.

   Disadvantages:

   1. Possibility of more errors.

2. **Calibration curve method:** A series of standard solutions are prepared from the stock solution. The appropriate filter is selected in the instrument, and air pressure of about 0.4-0.5kg /sq.cm is set in the instrument. Distilled water is sprayed into the flame, and the instrument is set to 0% flame intensity. Using the highest concentration of standard, 100% flame intensity is set. Then the flame intensity of other standard solutions is determined. A calibration curve is constructed with a concentration of the standard on the X axis and flame intensity on the Y-axis. The % flame intensity of the sample is recorded, and the concentration of the sample is found out by the interpolation of the graph.

3. **Standard addition method:** An equal amount of the sample solution is taken in different standard flasks, and increasing concentrations of standard solutions of the same element are added to each standard flasks. The flame intensity of all the solutions is recorded, and the calibration graph is constructed with a negative x-axis (sample concentration). The extrapolated line meets the negative x-axis to give the concentration of the sample. A typical calibration curve is shown in figure 4.8.

   Advantages:

   1. Low concentrations of the elements can be determined.

   2. Unexpected interfering substances will not affect the method.

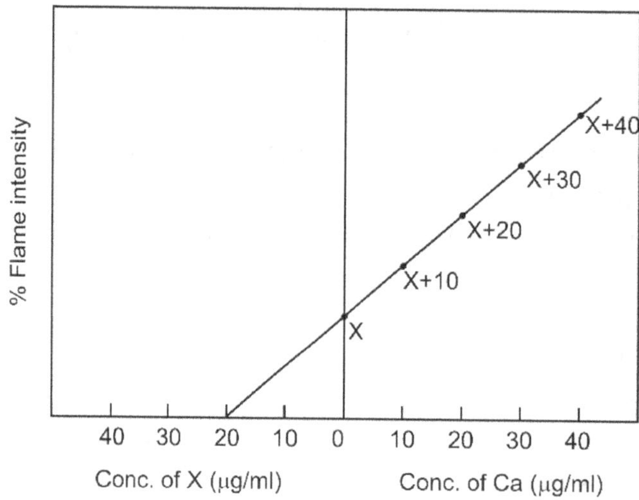

Fig. 4.8 Calibration curve in Standard addition Method.

4. **Internal standard method:** A known amount of the internal standard element is added to both the standard and sample solutions. The signal intensities of both sample and standard solutions are recorded. The ratio of the intensities is plotted against the concentration of the standard (calibration graph).The unknown concentration can be obtained from the graph by interpolation.

INTERFERENCES IN FLAME PHOTOMETRY

The presence of other materials in the sample causes some interferences in the recording of the spectrum. **Interference is a phenomenon that affects the measurement of the concentration of atoms in the ground state of an analyte.** The different types of interferences are

(i) **Physical interferences:** Background absorption, Spectral line interferences, matrix interference, viscosity and solvent interferences

(ii) **Chemical interferences:** Anionic interferences, vaporization interferences

(iii) **Scattering effects**

(iv) **Ionisation interferences**

(i) **Physical interferences**

1. **Background absorption:** The light absorption by the unvaporized solvent droplets in the flame causes background absorption. Absorption of unknown molecular species in flame and matrix interference (below 350nm) also causes background absorption.

   Correction: Using a deuterium lamp and grating monochromator minimizes these interferences.

2. **Spectral line interferences:** Occurs due to partial or complete overlap of the emission spectrum of two elements in the same wavelength. These interferences are more common in high flame temperatures.

   E.g., Iron 324.7 nm interferes with copper at 324.7 nm

Iron 285.2 nm interferes with Magnesium 285.2 nm

Aluminium interferes with emission lines of calcium and magnesium.

Correction:

1. Extraction of interfering material
2. Using calibration curves of solutions with the interfering element
3. Use of grating monochromators

3. **Matrix interference:** Sample matrix components other than the analyte react to form molecules to give this interference. Use of a blank to zero the instrument avoids this interference

   Changes in viscosity of the solution by a change in solvent or change in concentration causes matrix or bulk interference.

   Correction: Addition of an organic solvent decreases viscosity.

4. **Solvent interferences:** Metals in aqueous solutions give lower absorbance readings than the same metals in organic solutions.

(ii) **Chemical interferences:** When an analyte is not completely decomposed in flame to produce atoms, chemical interferences occur, leading to reduced absorbance of the analyte.

1. **Vaporisation interferences**

   There are two types in this interference

   (a) Chemical type
   (b) Physical type
   (a) **Chemical type** - occurs due to the incomplete dissociation and reaction with other ions in the sample.

      These interferences are classified as

   (i) Cation –Anion interferences
   (ii) Cation – cation interferences
   (iii) Oxide formation
   (i) **Cation – Anion interferences:** The intensity of emitted radiation is affected by the presence of certain anions like oxalate, phosphate, sulphate and aluminate.

      E.g., calcium in the presence of phosphate – calcium signal is suppressed.

   Correction:

   1. Extraction of anions
   2. Preparation of calibration graphs with a known concentration in the standard solutions (standard addition method).
   3. Choice of flame, burners, atomizer and additives

      E.g., Acetylene nitrous oxide flame for phosphates, sulphates, silicates, and aluminates which are thermally stable.

   4. Additives:
   (a) **Releasing agent:** Use of Ionization suppressants like Lanthanum/strontium to overcome interference due to Phosphates.
   (b) **Chelation or Masking Agents:** Use of EDTA to mask calcium in the presence of phosphate

(ii) Cation – Cation interferences: These interferences are caused by the presence of other cations. These interferences are neither spectral nor ionic in nature.

E.g., Aluminium interferes with calcium and magnesium

Sodium and potassium have cation–cation interference

Correction:

1. Extraction of interfering cations
2. Construction of calibration graph

(iii) Oxide formation interferences: These interferences arise due to the formation of stable oxides of metal ions due to the oxygen present in the flame. Thus, the emission is decreased. Metals like La, Al, and Ti give rise to oxides when aspirated into the flame.

Correction:

1. Use of high-temperature flames (Nitrous oxide-acetylene flames) to dissociate the oxides to give free atoms.
2. Oxygen free environment to produce atoms

(b) Physical type – High viscosity decreases intensity (interferes with atomisation.)

E.g., Dextrose, Sucrose.

This interference can be corrected by adding methanol to the sample, which enhances nebulisation and increases the sample concentration in flame leading to higher absorbance values.

(iii) Scattering effects: These effects also occur due to the higher concentration of the interfering element.

This effect can be eliminated by using a continuous light source (e.g., deuterium lamp) in addition to an HCL lamp.

Scattering occurs as a result of the particulate matter from the flame atomization scattering the incident radiation from the source. This can be corrected by using a blank during analysis.

(iv) Ionisation interferences: If the temperature of the flame is high, leading to ionization of the atoms, atomic emission intensity decreases. Ions have their own emission spectrum, which is different from atomic emission. This type of interference occurs only in I group elements (Na, K, Sr, etc.).

Correction: By adding excess (100-1000 µg) of easily ionisable salts like potassium, caesium, and strontium to the sample and standard solutions. These ions act as suppressants for the ionization of the analyte and enhance the emission of analyte molecules. These ions have lower ionization potential, and they get ionized preferentially over the sample analyte.

LIMITATIONS OF FLAME PHOTOMETRY

1. No information is obtained about the molecular form of metal present in the sample.
2. Inert gases cannot be detected by this method
3. Only liquid samples can be analysed. (solubility of sample in solvent)
4. The number of elements that can be analysed by flame photometry are limited.

CHAPTER 5

Atomic Absorption Spectroscopy

Atomic absorption spectroscopy is a study of absorption of a specific wavelength of radiation by the atoms of metals in the ground state. It is a method of elemental analysis. It is a highly versatile technique so that 60-70 elements, including common rare earth metals can be determined by direct methods. Many non-metals can be detected by indirect methods. It is a very sensitive technique.

PRINCIPLE

The solution containing the metallic salt sample is nebulized into the flame, fine droplets are formed. The thermal energy of the flame causes desolvation leaving the salt residue, which decomposes to give the neutral atoms. These neutral atoms absorb the specific wavelength of light emitted by the source, Hollow Cathode Lamp (HCL). For each element determination, the same element is used as cathode in the HCL source.

The intensity of light absorbed by the atoms is directly proportional to the concentration of the element, which is measured by the detectors. In atomic absorption spectroscopy, beer's law is obeyed over a wide concentration range.

Total amount of light absorbed $\upsilon = \pi e^2 \times Nf / mc$

Where,

  e - charge on the electron of mass, m

  c - speed of light

  N - total no of atoms that are absorbed at a frequency υ in the light path

  f - the ability of the atom to absorb at frequency υ

  e,m,π and c are constants.

  Thus, the total amount of light absorbed $=$ constant $\times N \times f$

This means that the amount of light absorbed is dependent on the number of atoms absorbing the light and their ability to absorb at that particular wavelength. The equation also shows that the amount of light absorbed is independent of the wavelength of absorption and temperature. The flame is used for atomizing the sample only, and the excitation of the atoms are brought about by the HCL source radiation.

Table 5.1 Difference between atomic absorption spectroscopy and flame emission spectroscopy.

Atomic absorption Spectroscopy	Flame emission spectroscopy
1. Measures the intensity of radiation absorbed by unexcited atoms of the analyte at the ground state.	Measures the intensity of radiation emitted by the excited state atoms.
2. Atomic absorption depends on the no. of unexcited atoms.	Flame spectroscopy depends on the no. of excited atoms.
3. Independent of flame temperature.	Temperature-dependent.
4. Nearly a linear relationship between absorption and concentration. Beer's law obeyed over a wide concentration range.	Not true with flame emission
5. A wide range of elements can be analyzed.	Only group IA, IIA elements can be analyzed.

The flow chart representing the principle of AAS is given

Sample aspirated into flame $\to$ formation of droplets $\to$ solid residue $\to$ neutral atoms formation $\to$ absorption of specific wavelength by neutral atoms from HCL $\to$ measurement of the intensity of absorbed radiation by photometric detectors.

Advantages of Atomic absorption spectroscopy (AAS) over flame emission spectroscopy
1. AAS is specific for the element under analysis and spectral interferences are rare.
2. Independent of flame temperature.
3. More elements can be determined. Sensitivity is more.

Disadvantages
1. A separate lamp is required for each determination
2. Elements giving rise to oxides in the flame cannot be determined. E.g., Al, Ti, W, Si
3. Anionic interferences are present.

INSTRUMENTATION

Components of the AAS spectrophotometer are
1. Radiation source
2. Nebulizers
3. Atomizer - Burners
4. Monochromator
5. Detector
6. Readout device

Table 5.2 Summary of Instrumentation.

Radiation source	Nebulizers	Atomizer - Burners	Monochromator	Detector
(i) Hollow cathode lamp (HCL) (ii) Electrodeless discharge lamp (EDL)	Pneumatic or ultrasonic nebulisers Suction, gravity-fed, controlled flow reflux nebulizers Angular and concentric nebulisers	1. Flame atomizers: Total consumption and premixed burners 2. Non-flame atomizers: Graphite tube atomizer carbon filament atom reservoirs	Filter Prism Grating	Phototubes Photomultiplier tube

1. **Radiation sources:** The two most commonly used radiation sources are
   (i) Hollow cathode lamp (HCL)
   (ii) Electrodeless discharge lamp (EDL)
   (i) Hollow cathode lamp: It is the most common AAS source.

   It contains a tungsten anode and a hollow cylindrical cathode.

   The cathode is made up of the element to be determined. Thus, a separate lamp is required for the determination of each element.

   The cathode may also be made up of alloy for the determination of the elements present in it. Then it is called a multi-element lamp.

   The anode and the cathode are sealed in a glass cylinder filled with argon or neon gas at low pressure. The opening of the glass cylinder has a silica/quartz window, transparent to the emitted radiation.

   The potential applied between the electrodes causes a current in the mA range to flow through. The inert gas gets ionized at the anode, and these ions of gas are attracted to the cathode with high velocity. The impact with the cathode vaporizes the atoms of the cathode element. The vaporized atoms collide with the inert gas atoms or molecules and get excited. These excited atoms emit their characteristic radiation, and this emitted radiation is absorbed by the neutral atoms of the same element in the sprayed sample solution.

   Advantages
   1. It gives a bandwidth of 0.001 to 0.01 nm

2. The line width provided is less than the absorption line width of the element to be determined.

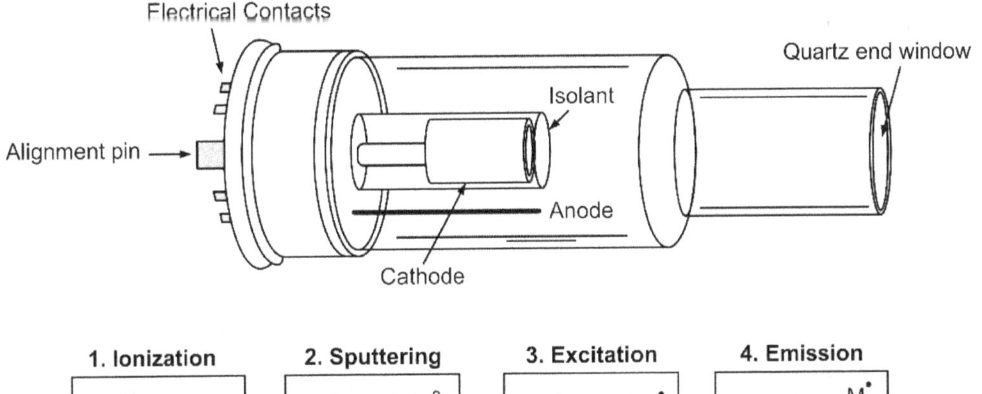

Fig. 5.1 Hollow cathode lamp.

Disadvantages

1. Finite lifetime due to the depletion of cathode
2. Adsorption of gas atoms on the inner surface of the glass cylinder.
3. For the determination of every element, a separate lamp should be used.

(ii) Electrodeless discharge lamp: This lamp consists of a small amount of the element or a salt of the element sealed in a quartz bulb with an inert gas atmosphere. The bulb is placed inside a small RF generator. When an RF field of sufficient power is applied, the inert gas is ionized. The coupled energy vaporizes the element and excites the atoms inside the bulb, resulting in the emission of the characteristic spectrum. EDC can be used to detect elements including Sb, As, Bi, Cd, Cs, Ge, Pb, Hg, P, K, Rb, etc.

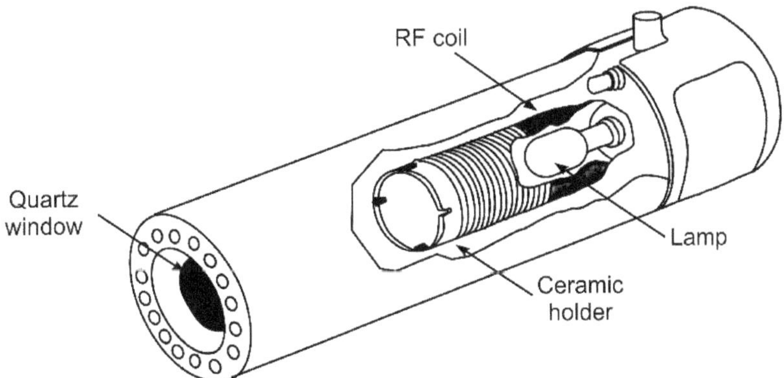

Fig. 5.2 Electrodeless discharge lamp.

Chopper: The function of the chopper is to create fluctuation in the output of the source. A chopper (rotating wheel) is placed between the hollow cathode lamp and the flame. This breaks the steady light from the lamp into pulsating light. This fluctuating light produces a pulsating current in photometric detectors. The light emitted by the flame produces a steady current. In the instrument, only the pulsating current is amplified and recorded. Thus, the interference due to light emitted by the flame is eliminated.

2. **Nebulizers:** The conversion of the liquid sample into small droplets is called nebulization. The liquid sample is drawn up the capillary by a high-velocity gas stream. A fine aerosol spray is mixed with fuel and oxidant gas thoroughly and then introduced into the flame.

3. **Atomizers:** In order to analyze the samples by AAS, the sample should be in an atomic state. This is achieved by the breaking down of the molecules into atoms. There are two types of atomizers.

   (i) Flame atomizers

   (ii) Non-flame atomizers.

   (i) **Flame atomizers:** Burners are used as flame atomizers

   The burners are used to spray the sample solution into the flame after mixing with fuel and oxidant. There are two types of burners, Total consumption and premixed burners.

   Total consumption burner: In these burners, hydrogen and oxygen gases are used as fuel and oxidant gases respectively. The liquid sample is aspirated into the flame through a capillary tube. The oxygen and hydrogen enter through the side tubes and are burnt at the top of the burner to produce a flame. As soon as the sample enters the flame, it is vaporized, atomized and then excited.

   It is called a total consumption burner since the entire sample entering the capillary is consumed.

   Advantages

   1. Simple design
   2. Entire sample consumed
   3. No explosive hazard.

   Disadvantages

   1. Large droplets not decomposed completely
   2. Non- uniform and non-homogenous flame
   3. Fluctuations in flame intensity
   4. Viscosity of the sample affects the rate of sample introduction
   5. Short path for the solute decreases sensitivity

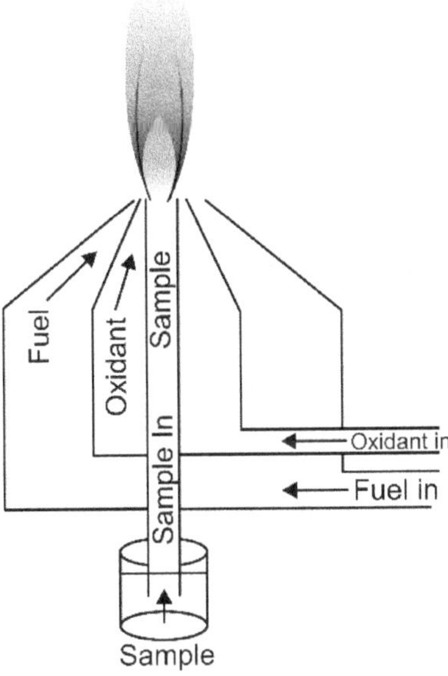

Fig. 5.3 Total consumption burner.

Premix laminar flow burners –most widely used burner: In this burner, the sample fuel and oxidant gas are mixed thoroughly and then introduced into the flame. The flow of the gases is laminar and not turbulent as in the total consumption burner. Only a small fraction of the sample droplets reach the flame. Easy decomposition of the sample takes place, leading to efficient atomization. Large droplets are drained off the outlet.

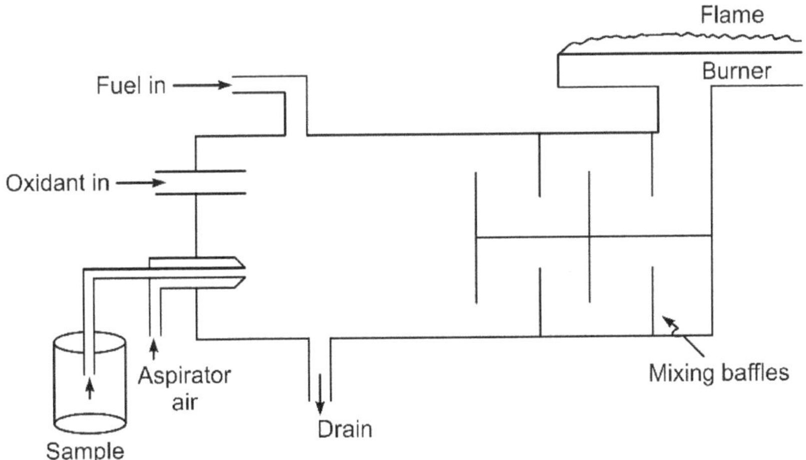

Fig. 5.4 Premix laminar flow burners.

Advantages

1. Noiseless, non-turbulent and stable flame.

2. Efficient atomization

3. Clogging free

4. Long path provides sensitivity

Disadvantages

1. Only a small quantity of the sample reaches the flame

2. Rate of sample introduction is slow.

(ii) Non-flame atomizers: E.g. of these types of atomizers are graphite tube atomizers and carbon filament atom reservoirs. A similar process is used in both the atomizers

Graphite tube atomizer: A small quantity of the sample (2-30μl) is loaded in the sample compartment. Here the sample undergoes desolvation, and it decomposes to form the atoms by electrical heating. Thus, the free atoms are vaporized and introduced into the optical light path, and the absorption of light is measured.

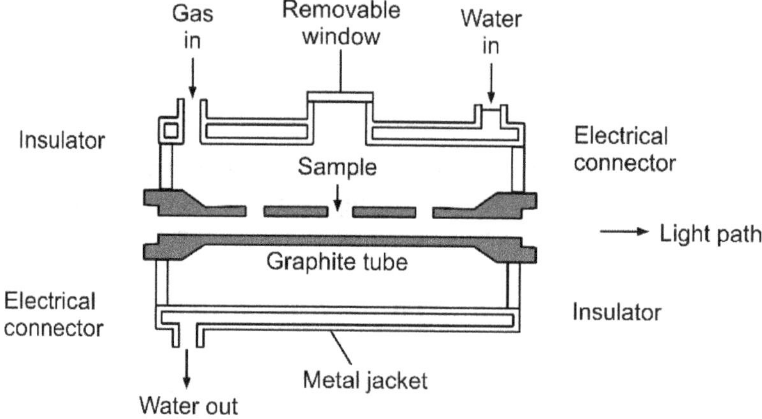

Fig. 5.5 Graphite tube atomizer.

4. **Monochromators:** The most common monochromators are prism and gratings. These monochromators are explained in detail in the UV Visible spectroscopy chapter. It has to select the wavelength emitted by the element, excluding the emission by the gas in the HCL lamp or the background signal of flame.

5. **Detector:** The radiations absorbed by the elements are in the UV-visible range. The most suitable detector is a photomultiplier tube. In some instruments, phototubes are used.

6. **Readout device:** Readout devices display the absorption spectrum and the absorbance at a specified wavelength.

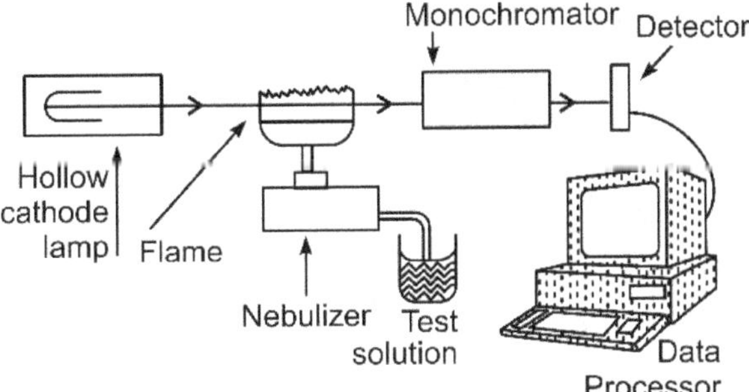

Fig. 5.6 Schematic diagram of atomic absorption spectrophotometer.

Two types of AA Spectrophotometers are available. They are
1. Single beam atomic absorption spectrometer
2. Double beam atomic absorption spectrometer

1. **Single beam atomic absorption spectrometer:** The light source is placed ahead of the flame with a chopper in between the light source and flame; when there is no sample in flame, the detector records the full intensity of the source emission. When the sample is introduced into the flame, there is a reduction in the emission due to absorption by the sample atoms. This output is recorded by the detector.

Disadvantages

Low stability.

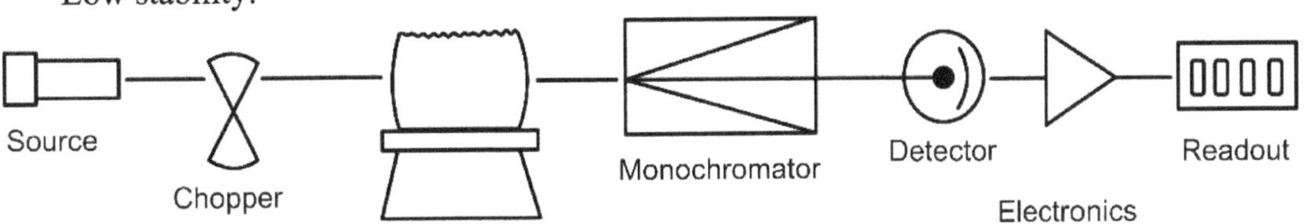

Fig. 5.7 Schematic diagram of Single beam atomic absorption spectrophotometer.

2. **Double beam atomic absorption spectrometer:** The chopped beam from the source (hollow cathode lamp) is divided into two parts. One part passes through the flame, and the other is the reference beam not passing through the flame. The two beams are recombined and reach the detector through a monochromator. The detector output is displayed in the readout system.

Advantages

1. Avoids fluctuations in the conditions of the flame
2. Viscosity, temperature and spraying rate effects are eliminated.
3. No change in detector sensitivity with time.

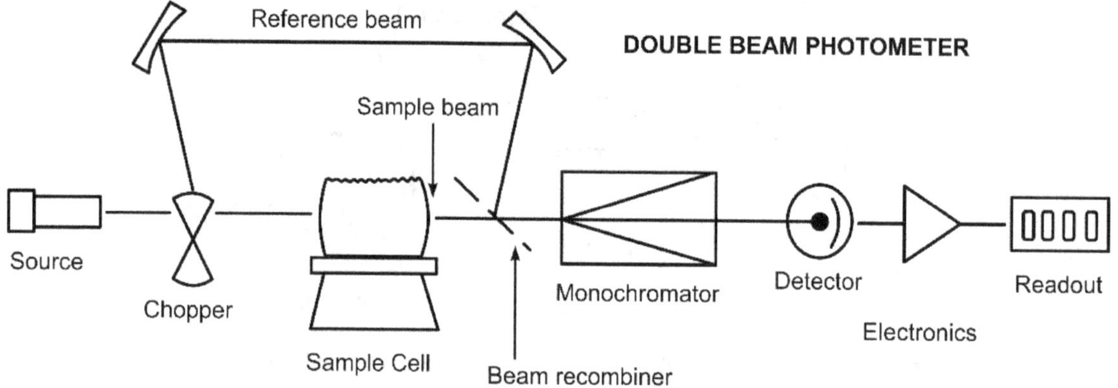

Fig. 5.8 Schematic diagram of Double beam atomic absorption spectrophotometer.

INTERFERENCES

Interference is a phenomenon that affects the measurement of the concentration of atoms in the ground state of an analyte.

The interference is less in AAS when compared to flame photometry. It is also free from cationic interferences due to the absorption of sharp resonance lines from HCL. The different types of interferences are

(i) Physical interferences: Background absorption, Spectral line interferences, matrix interference, viscosity and solvent interferences

(ii) Chemical interferences: Anionic interferences, vaporization interferences

(iii) Scattering effects

(iv) Ionisation interferences

(i) Physical interferences

    1. Background absorption: The light absorption by the unvaporized solvent droplets in the flame causes background absorption. Absorption of unknown molecular species in flame and matrix interference (below 350nm) also causes background absorption.

    Correction: The use of a deuterium lamp and grating monochromator minimizes these interferences.

    2. Spectral line interferences: Occurs due to partial or complete overlap of the emission spectrum of two elements in the same wavelength. These interferences are more common in high flame temperatures.

    E.g., Iron 324.7 nm interferes with copper at 324.7 nm

    Iron 285.2 nm interferes with Magnesium 285.2 nm

    Aluminium interferes with emission lines of calcium and magnesium.

    Correction:

    1. Extraction of interfering material

2. Using calibration curves of solutions with the interfering element

3. Use of grating monochromators

3. Matrix interference: Sample matrix components other than the analyte react to form molecules to give this interference. Use of a blank to zero the instrument avoids this interference

Changes in viscosity of the solution by a change in solvent or change in concentration cause matrix or bulk interference.

Correction: addition of an organic solvent decreases viscosity.

4. Solvent interferences: Metals in aqueous solutions give lower absorbance readings than the same metals in organic solutions.

(ii) **Chemical interferences:** When an analyte is not completely decomposed in flame to produce atoms, chemical interferences occur, leading to reduced absorbance of the analyte.

1. Vaporization interferences: There are two types in this interference

   (a) Chemical type

   (b) Physical type

   (a) **Chemical type** - occurs due to the incomplete dissociation and reaction with other ions in the sample.

These interferences are classified as

   (i) Cation –Anion interferences

   (ii) Cation – cation interferences

   (iii) Oxide formation

   (i) **Cation – Anion interferences:** The intensity of emitted radiation is affected by the presence of certain anions like oxalate, Phosphate, sulphate and aluminate.

   E.g., calcium in the presence of Phosphate – calcium signal is suppressed.

   Correction:

   1. Extraction of anions

   2. Preparation of calibration graphs with a known concentration in the standard solutions (standard addition method).

   3. Choice of flame, burners, atomizer and additives

      E.g., Acetylene nitrous oxide flame for phosphates, sulphates, silicates, and aluminates which are thermally stable.

   4. Additives:

   (i) Releasing agent: Ionisation suppressants like lanthanum/strontium to overcome interference due to Phosphates.

   (ii) Chelation or Masking Agents: Use of EDTA to mask calcium in the presence of Phosphate

(ii) Cation – cation interferences: These interferences are caused by the presence of other cations. These interferences are neither spectral nor ionic in nature.

E.g., Aluminium interferes with calcium and magnesium

Sodium and potassium have cation–cation interference

Correction:

1. Extraction of interfering cations
2. Construction of calibration graph

(iii) Oxide formation interferences: These interferences arise due to the formation of stable oxides of metal ions due to the oxygen present in the flame. Thus the emission is decreased. Metals like La, Al, and Ti give rise to oxides when aspirated into the flame.

Correction:

1. Use high-temperature flames (Nitrous oxide-acetylene flames) to dissociate the oxides and give free atoms.
2. Oxygen free environment to produce atoms

(b) Physical type – High viscosity decreases intensity (interferes with atomization.)

E.g., Dextrose, Sucrose.

This interference can be corrected by adding methanol to the sample, which enhances nebulization and increases the sample concentration in flame, leading to higher absorbance values.

(iii) Scattering effect: These effects also occur due to the higher concentration of the interfering element.

This effect can be eliminated by using a continuous light source (e.g., deuterium lamp) in addition to an HCL lamp.

Scattering occurs as a result of the particulate matter from the flame atomization scattering the incident radiation from the source. This can be corrected by using a blank during analysis.

(iv) Ionization interferences: If the temperature of the flame is high, leading to ionization of the atoms, atomic emission intensity decreases. Ions have their own emission spectrum, which is different from atomic emission. This type of interference occurs only in I group elements (Na, K, Sr, etc.).

Correction: By adding excess (100-1000 µg) of easily ionizable salts like potassium, caesium, and strontium to the sample and standard solutions. These ions act as suppressants for the ionization of the analyte and enhance the emission of analyte molecules. These ions have lower ionization potential, and they get ionized preferentially over the sample analyte.

APPLICATIONS OF AAS

1. **Qualitative analysis:** Since different HCL source has to be used for each element, AAS is not used in qualitative analysis.

2. **Quantitative analysis:** The calibration curve method is used. Serial dilutions of standard solutions are prepared, and their absorbances are measured.

3. A calibration curve is constructed with the concentration of elements and absorbance.

4. The concentration of the extrapolation of the graph. Beer's law is obeyed over a wide range of concentrations. Sample concentration upto 0.001 ppm can be analyzed quantitatively in AAS.

5. The other applications are

(i)   Analysis of trace elements in biological fluids. (e.g., Blood, wine, etc.)

(ii) Estimation of elements like Ca, Ni, and Zn in food products.

(iii) In the pharmaceutical industry, it is used to estimate zinc in zinc insulin injection, Mercury in thiomersal injection.

(iv) Analysis of elements in soil samples, water, effluents, ceramics, etc.

(v)   Determination of lead in $CaCO_3$, petrol etc.

CHAPTER 6

Nepheloturbidimetry

Nepheloturbidimetry is based on the scattering of light by non-transparent particles suspended in a solution. When light passes through a suspension, a part of the incident light is absorbed, reflected, refracted, and the remaining part is transmitted.

Measurement of the intensity of **transmitted radiation** as a function of the concentration of suspended particles is called turbidimetric analysis. The measurement is made at 180°angle.

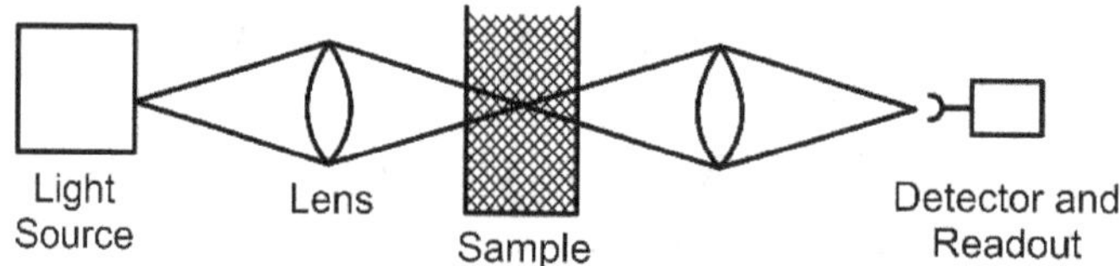

Fig. 6.1 Schematic diagram of Turbidimetric analysis.

Measurement of the intensity of **scattered light** as a function of the concentration of suspended particles is called Nephelometric analysis, where the measurement is made at 90° angle.

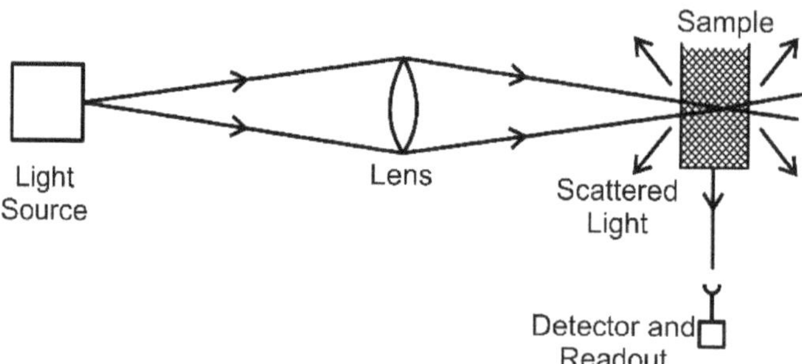

Fig. 6.2 Schematic diagram of Nephelometric analysis.

PRINCIPLE

Nephelometry: The principle behind nephelometry is that there is a uniform scattering of light in low concentrations of a suspension. In this condition, the **intensity of the scattered light is directly proportional to the concentration.** The scattered light is measured at 90°. If the particles are large, scattering at 90° is less, so measurements can be made at angles like 5 to 20° or 45°.

Turbidimetry: At high concentrations of the suspensions, scattering is not uniform, and the light gets scattered in all directions. Thus it is difficult to measure the scattered light. To overcome this difficulty, the intensity of the transmitted light is measured at 180°.The principle behind turbidimetric measurements is that the **intensity of transmitted radiation is inversely proportional to concentration.**

THEORY

REFLECTION VS SCATTERING

If the dimensions of the suspended particle are more larger in magnitude than the incident light wavelength, reflection takes place.

Particles of equal or smaller magnitude than the incident wavelength will cause scattering.

This difference plays an important role in Nepheloturbidimetry.

In Nephelometry, the suspended particles should be smaller in size when compared to wavelength so that scattering will be predominant than transmission. Smaller particles give maximum scattering intensity at 90° and hence the **detector is placed at 90° in nephelometer.**

If larger particles are there, then more scattering occurs in angles other than 90° and the measurements are made at angles in the region of 5° to 20° or 45°

The particle size in nephelometry should be in the range of 0.1 to 1μm.

In **turbidimetry,** only the **transmitted radiation is measured,** and so the particle **size does** not play an important role. But the linear relationship between absorbance and concentration is not maintained with larger particles.

FACTORS AFFECTING NEPHELOTURBIDIMETRIC MEASUREMENTS

Light scattering is dependent on

1. Concentration
2. Particle geometry
3. Wavelength
4. Refractive index difference

1. **Concentration:** In turbidimetry,
   Transmittance $T = I/I_0$ ....(1) where,
   I_0 – the intensity of incident light
   I – the intensity of transmitted light
   The transmittance T and concentration C are correlated as
   $$S = \log I_0/I = kbc \ \ldots\ldots \ (2) \quad \text{where}$$
   S – Turbidence (due to scattering)

b – Pathlength

k – constant (turbidity coefficient)

The value of k depends on particle size, shape, wavelength and refractive indices of dispersed and dispersion media.

Equation ② holds good only for dilute solutions. In dilute solutions, Turbidence is directly proportional to concentration.

The equation for nephelometry is

$$I_S = k_S I_0 C \quad \ldots\ldots \text{③} \quad \text{where}$$

k_S – is a constant,

I_S – intensity of scattered radiation

I_0 – the intensity of incident radiation,

C-Concentration of suspended particles.

In both methods, quantitative estimation is carried out by preparing a calibration/working curve.

Turbidimetry – S (turbidence) plotted against known concentrations of the analyte.

Nephelometry – log (I_0/I_s) against concentration

Unknown concentrations are determined from the graphs in both methods.

2. **Particle Geometry:** The fraction of light scattered at any angle depends upon the size and shape of the particles. The extent of scattering(s) is directly proportional to the square of the effective radius of the particle. To control the particle size and shape, samples and standards must be prepared under identical conditions.

The conditions to be monitored are

(i) The concentration of the reactants

(ii) Temperature.

(iii) Agitation and pH

(iv) Order of mixing reagents

(v) Presence of other inert electrolytes.

3. **Wavelength:** The intensity of scattered radiation depends upon the wavelength of the incident light. The analyte should not absorb in the selected wavelength. Generally, in nepheloturbidimetry, white light is used.

4. **Refractive index difference:** There should be an appreciable difference in the refractive index of the dispersed and dispersion medium to get good results.

Table 6.1 Comparison between Nephelometry and Fluorimetry.

Nephelometry	Fluorimetry
Measurement of scattered light at 90°	Measurement of emitted light at 90°
The intensity of scattered radiation is measured	The intensity of emitted radiation is measured
Scattered and incident radiation have the same wavelength (in elastic)	Emitted radiation occurs at a longer wavelength than incident radiation (elastic)

Table 6.2 Comparison between Colorimetry and Turbidimetry.

Turbidimetry	Colorimetry
Transmitted light measured at 180°	Transmitted light measured at 180°
The decrease in intensity of light is due to the scattering of radiation	The decrease in intensity of light is due to the absorption of radiation

CHOICE BETWEEN NEPHELOMETRY AND TURBIDIMETRY

The choice of the method depends on the concentration of the suspension

Low concentration of suspension - Nephelometry is the choice

In low concentrations, scattered light is less and can be measured at 90° easily.

High concentration of suspension – Turbidimetry is the choice

In high concentration, scattered light is more. Hence it is easy to measure the transmitted light at 180°.

INSTRUMENTATION

Separate instruments like Nephelometer and turbidimeter are available. Combination instruments like Nepheloturbidimeter are most commonly used.

The components are

1. Source of light
2. Filters and monochromators
3. Sample cells
4. Detectors

Table 6.3 The components of nepheloturbidimeter.

Source of light	Filters and monochromators	Sample cells	Detectors
1. Mercury arc lamp- monochromatic radiation 2. Tungston lamp- Polychromatic light	For polychromatic light - No filters /monochromators. Monochromatic light-filters/ Monochromators needed. **Turbidimeter**-blue filter or 530nm Nephelometer-visible filter (secondary filter)	Glass cuvettes Cylindrical, rectangular, semi-octagonal	**Nepheloturbidimeter-** phototubes, photovoltaic cells and PMT **Turbidimeters –** Phototubes **Nephelometers** -PMT

1. Source of light: For determining the concentration of the material, a polychromatic source like, tungsten lamp is used. For monochromatic radiation, a Mercury arc lamp is used. This reduces the absorption of light since we measure only the scattering of light in nepheloturbidimetry.

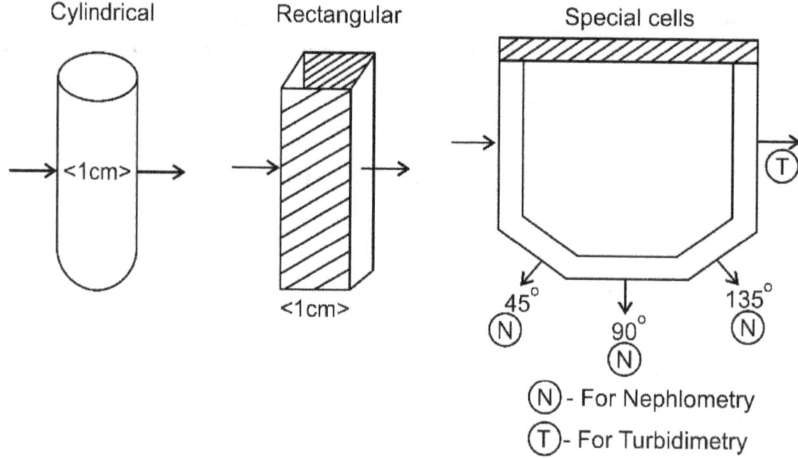

Fig. 6.3 Different types of sample cells used in nepheloturbidimetry.

2. **Filters and Monochromators:** When polychromatic light (white light) is used, filters and monochromators are not required. Only when monochromatic light is needed, filters and monochromators are used.

   A Blue filter or 530nm is used in the turbidimeter. The visible filter is used as a secondary filter in the nephelometer.

3. **Sample Cells:** The cells used may be cylindrical or rectangular cells with a 1cm path length. The material of construction is glass. The walls of the cells through which the light beams are not passing are painted dull black to absorb unwanted radiation and reduce the stray radiation. The semi-octagonal cells are used to measure the scattered light at different angles like 45°,90°,135° and 180°.

4. **Detectors:** In nepheloturbidimeters, photometric detectors like phototubes, photovoltaic cells and Phototubes are used. In a turbidimeter, Phototubes are used. In nephelometers, PMT is used as the detector and mounted at 90°.PMT detectors are used in nephelometers because the intensity of scattered radiations is weak.

INSTRUMENTS

NEPHELOMETER

The Source of light is a tungsten lamp. The sample cell is placed above the source, and light passes through the sample. The particles of the sample scatter the light which is reflected by the mirror to the photovoltaic cell. It is a simple, inexpensive model. Easy to operate. It has reasonable precision and accuracy.

A fluorimeter can be modified as a nephelometer by using a visible filter as a secondary filter.

TURBIDIMETER

The schematic diagram of the turbidimeter is shown in Fig. 6.5.

Light from the source passes through the filter to the sample cell. The light transmitted after passing through the sample cell is measured by the detector. A colorimeter can be modified as a turbidimeter by using a 530nm or blue filter.

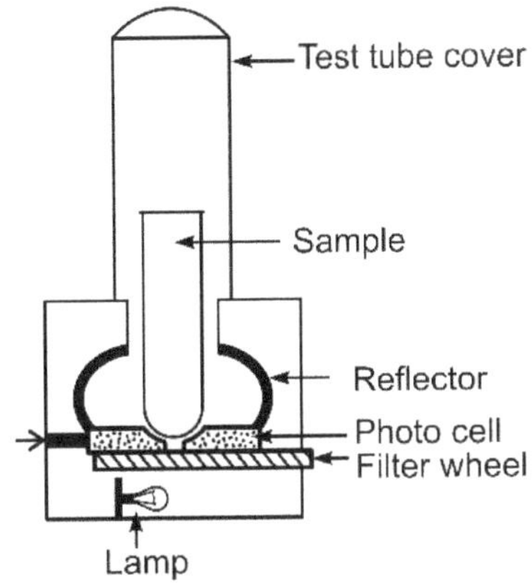

Fig. 6.4 Nephelometer.

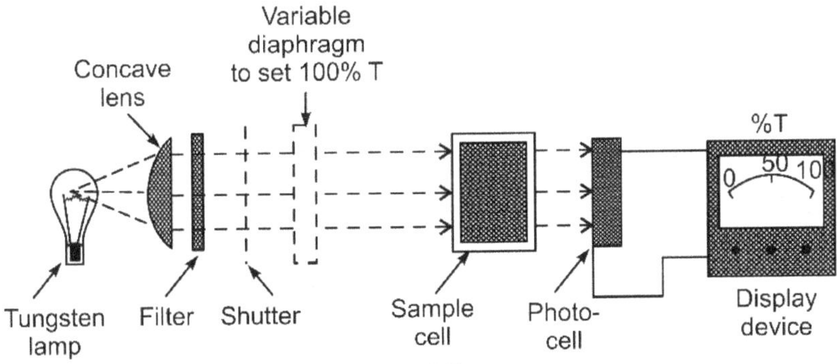

Fig. 6.5 Turbidimeter.

NEPHELOTURBIDIMETERS

The nepheloturbidimeters are a combination of both nephelometer and turbidimeter. The nepheloturbidimeter has two detectors. One of the detectors is placed at 90° to measure the scattered light and the other detector at 180° to measure the transmitted light. The response is the ratio of two detectors, which is displayed as NTU units (Nepheloturbidimetric units),which is proportional to the concentration of the suspension.

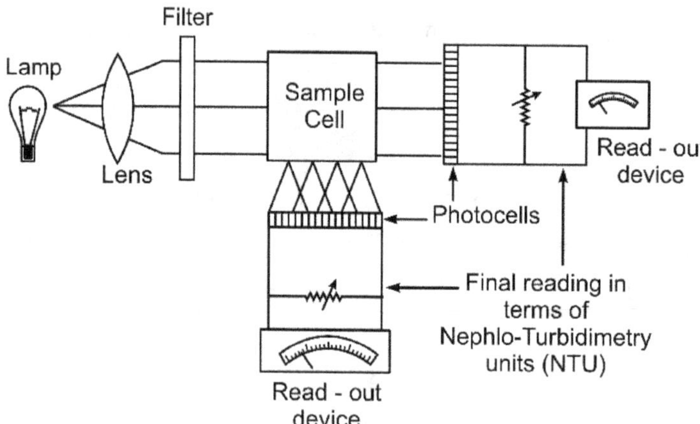

Fig. 6.6 Nepheloturbidimeter.

APPLICATIONS

1. **Inorganic analysis:** Inorganic elements like sulphate, carbonate, chloride, fluoride, cyanide etc., can be estimated using nepheloturbidimetry. The elements are precipitated using specific reagents and the turbidity produced is measured. Some of the elements and their precipitating reagents are given in table 6.4

Table 6.4 Some of the elements and their precipitating agents.

Element	Precipitating reagent	Precipitate formed
Sulphate	Barium chloride	Barium sulphate
Chloride	Silver nitrate	Silver chloride
Carbonate	Barium chloride	Barium carbonate
Fluoride	Calcium salts	Calcium fluoride
Cyanide	Silver salts	Silver cyanide
Calcium	Oxalate salts	Calcium oxalate
Zinc	Potassium ferrocyanide	Zinc ferrocyanide
Ammonia	Nessler's reagent	Ammonium mercuric iodide
Phosphorous	Strychnine molybdate	Phosphorous molybdate

2. **Organic analysis:** The turbidimetric analysis is used to find turbidity in sugar products and the clarity of citrus juices.

3. **Biochemical analysis:** Turbidimetry is used in the assay of antibiotics, vitamins and amino acids. Nephelometry is used to determine globulins in serum and plasma, yeast and glycogen.

4. **Air and water pollution:** Nepheloturbidimetry is used in the monitoring of air and water pollution. Dust and smoke in the air and turbidity in water are monitored.

5. **Determination of carbon dioxide (CO_2):** Carbon dioxide gas passed through barium chloride solution; the turbidity produced due to the formation of barium carbonate is determined by Nepheloturbidimetry.

6. **Water analysis:** The concentration of various ions in water is determined using precipitating reagents

7. **Quantitative analysis of ions (ppm level):** Serial dilutions of standard solution are prepared and treated with reagents to produce turbidity. The sample is also treated similarly. The reading is measured in NTU.A calibration curve is constructed using standard readings, and the concentration of the unknown solution is determined from the graph.

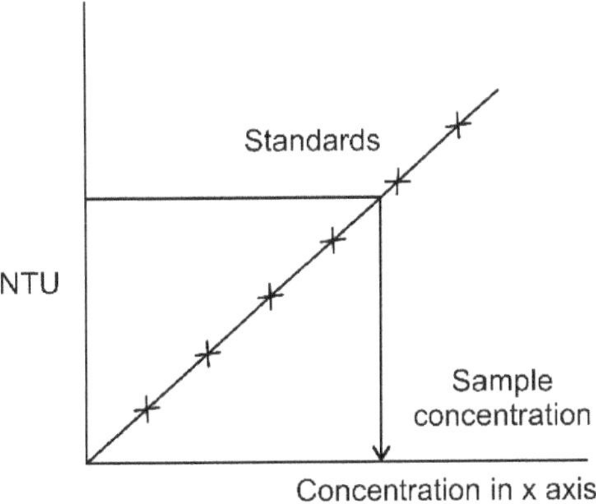

Fig. 6.7 Calibration graph.

8. **Miscellaneous:** Nepheloturbidimetry is used in water treatment plants, sewage works, power generating plants, breweries, bottling industry, petroleum refineries, paper industry etc.

9. **Turbidimetric titrations:** These titrations are similar to spectrophotometric titrations.NTU is plotted against the volume of the titrant in the titration curve. The titrant reacts with titrand to give a turbid product. The point of inflection in the graph gives the endpoint.

E.g., Titration of fluoride with calcium

   Titration of bromide with silver

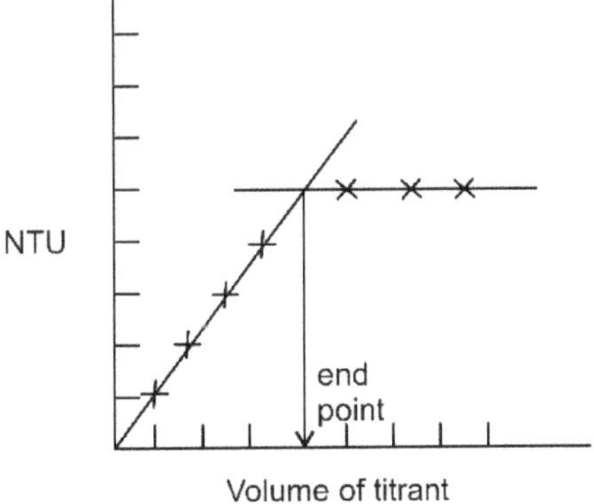

Fig. 6.8 Turbidimetric titration curves.

SECTION – II: Chromatography

CHAPTER 7

Introduction to Chromatography

INTRODUCTION

Chromatography is a separation technique used to separate a mixture of various components into individual components using equilibrated distribution between two phases (stationary phase and mobile phase). The basis of chromatography is the rate differences in the movement of the components of a mixture through the stationary phase (porous medium) against the mobile phase (solvent or gas).

The steps involved in chromatographic separation are

1. Adsorption or retention of the analyte components in the stationary phase.
2. Isolation of the adsorbed/retained components by the mobile phase.
3. Elution of the isolated substances by the flow of the mobile phase.
4. Qualitative and quantitative estimation of eluted substances.

TYPES OF CHROMATOGRAPHY

Different methods of classification are available for chromatography. The classifications are

1. **Based on the nature of the stationary and mobile phase:** Depending on the type of the stationary and mobile phases used, the chromatographic methods can be classified as given below.

Table 7.1 Classification of chromatography based on stationary and mobile phase.

Mobile Phase	Stationary Phase	Name of the Technique
Gas	Solid	Gas Solid Chromatography (GSC)
Gas	Liquid	Gas Liquid Chromatography (GLC)
Liquid	Solid	Thin Layer Chromatography (TLC) Adsorption Column Chromatography High Performance Liquid Chromatography (HPLC)
Liquid	Solid matrix(Gel)	Gel Permeation Chromatography
Liquid	Liquid	Paper Partition Chromatography Column Partition Chromatography
Liquid	Solid-resin	Ion Exchange Chromatography

2. **Based on the principle of separation:** The principle of separation in chromatography can be adsorption or partition

   (i)   Adsorption chromatography

   (ii) Partition chromatography

   (i) **Adsorption chromatography:** The separation of components in adsorption chromatography is based on the relative affinities of the components towards the stationary phase. The principle of separation is called adsorption, which is a surface phenomenon. The mixture (adsorbate), which is dissolved in the mobile phase, moves through the stationary phase (adsorbent) of the column. The component with more affinity for the stationary phase travels slower, and the component with more affinity for the mobile phase travels faster down the column. Thus the components are separated based on their affinity difference for the stationary phase. Under defined conditions, no two components will have the same affinity for the given set of the stationary, mobile phase.

   If the stationary phase is solid, then the principle of separation is adsorption

   E.g., GSC, TLC, column adsorption chromatography, HPLC

   (ii) **Partition chromatography:** In this mode of chromatography, the separation of components is based on the partition coefficients of the individual components. Partition coefficient (P) is the ratio of concentrations of a compound in a mixture of two immiscible solvents at equilibrium. The solute gets distributed between the two immiscible liquids based on its partition coefficient. The more soluble component in the stationary phase travels slower than the component that is soluble in the mobile phase. The components are separated based on their differences in partition coefficient since no two components will have the same partition coefficient for the given set of conditions, including the stationary and mobile phases.

   The stationary phase is a liquid coated as a thin film over solid support.

   If the stationary phase is a liquid, then the principle of separation is partition.

   E.g., Gas-liquid chromatography, paper partition chromatography, column partition chromatography etc.

3. **Based on the modes of chromatography:** Modes of chromatography depend upon the polarities of the stationary and mobile phase. The interaction between solute, stationary and mobile phases are,

   (i)   Polar-Polar interactions

         Polar solute dissolves in a polar solvent.

   (ii) Non –polar- Non-polar interactions

         Non-polar solute dissolves in Non-polar solvent

         These interactions are based on "like dissolves like"

   (iii) Polar – Non- Polar interactions are very weak.

         In this classification, two modes of separation are available.

         (a) **Normal Phase chromatographic mode**

         (b) **Reverse phase chromatographic mode**

(a) **Normal Phase Chromatography**

Stationary phase is polar (silica gel)

Mobile phase is Non-polar

Polar components interact more with the polar stationary phase and are retained in the column for a longer time. Non-polar components exhibit the least interaction with the stationary phase and are eluted fast. This mode is **not much** used in **pharmaceutical applications** because the nature of most **drug molecules are polar** and are **retained more**, making the elution time longer.

(b) **Reverse phase chromatography**

Stationary phase is Non -polar (ODS- C_{18} or C_8 etc.)

Mobile phase is Polar

In this mode, polar components are eluted faster, and the non-polar components are retained for a longer time in the stationary phase. This is the most **commonly used** technique in **pharmaceutical analysis**.

The distinction between reverse phase and Normal phase mode is given in Table 7.2.

Table 7.2 Distinction between reverse phase and Normal phase mode.

Parameter	Normal Phase	Reverse Phase
Stationary phase	Polar	Non-polar
Mobile phase	Non-polar	Polar
Less retained (first eluted compound)	Non-polar	Polar
More retained (last eluted compound)	Polar	Non-polar
E.g. of stationary phase	Silica gel	ODS(C_{18}),C_8,C_4-bonded phases

4. Other types of chromatographic techniques

1. **Ion exchange chromatography:** The principle involved is the reversible exchange of ions. The stationary phase used is an ion exchange resin. Anion exchange resin is used to separate anions, and cation exchange resin is used to separate cations. Acids, alkalis and buffers are used as mobile phase.

2. **Ion–pair chromatography:** In this technique, ion-pairing agents like octane sulphonic acid, Tetraethyl ammonium hydroxide are used to convert a reversed-phase column into a temporary ion exchange column.

3. **Size exclusion /gel permeation chromatography:** This technique is also called as Gel filtration or Gel permeation chromatography. The stationary phase is a porous matrix made up of porous materials like cross-linked polystyrene, cross-linked dextrans, polyacrylamide gels, agarose gels, etc. The separation is based on the analyte molecular sizes since the gel behaves like a molecular sieve. This technique is used to separate proteins, polysaccharides, enzymes, and synthetic polymers. The solvents used are aqueous buffers or organic solvents.

4. **Chiral chromatography:** Chiral chromatography is used to separate enantiomers (dextro and levo form) using either chiral stationary or mobile phase.

Theories of chromatography

There are two theories of chromatography which deal with the rate of solute migration and peak development in the chromatogram.

(i) Plate theory
(ii) Rate theory

(i) **Plate theory:** This theory was developed by Martin and Synge in 1941. In plate theory, the column is divided into a number of adjacent imaginary segments called theoretical plates. In each theoretical plate, a complete equilibration of the solute occurs between the stationary and mobile phases.

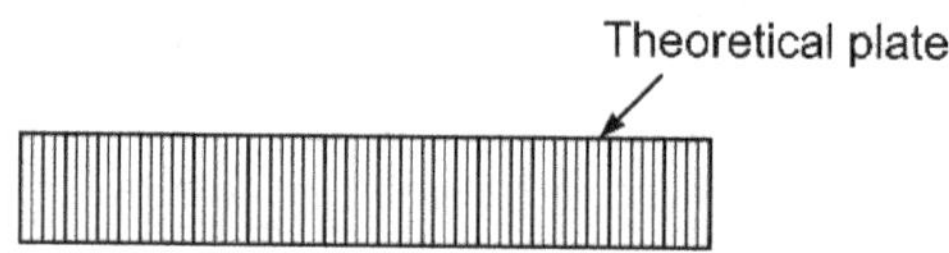

Fig. 7.1 Column showing imaginary theoretical plates.

The equal interaction of the solute components of the mixture with the mobile phase and stationary phase leads to the retention of the molecules (slow rate of movement).

Each solute molecule interacts with the mobile phase and stationary phase in a different manner and they will have different rates of migration leading to the separation of the components.

Significance of theoretical plates

Efficiency (good separation of the components) of the column increases with

1. Increase in the number of theoretical plates. (N)
2. Decrease in HETP (Height equivalent to theoretical plates)

$H = L/N$ where

L - Length of the column
N - Number of plates
H - HETP or Height of the theoretical plates.

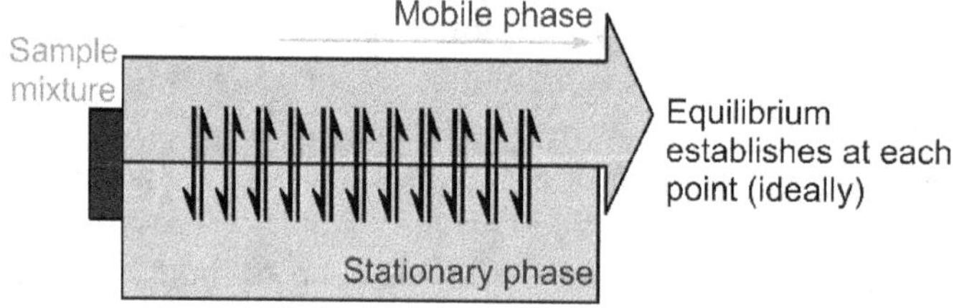

Fig. 7.2 Diagrammatic representation of Theoretical plates.

HETP is defined as the height of the theoretical plate in which a complete equilibration of the solute occurs between the stationary and mobile phase.

The molecules of the mixture interact with the molecules of the stationary and mobile phase leading to retardation of the molecular movement. Each molecule interacts differently with the mobile and stationary phase due to the **difference in their distribution coefficients and this causes net difference in rates of migration**

(ii) **Rate theory:** This theory was proposed by Van Deemter in the year 1956. This theory describes the processes that are working inside the column. The time taken for the equilibration of the solute between the stationary phase and the mobile phase is taken into account in this theory. The resulting band shape or chromatographic peak is affected by various factors that lead to band broadening.

Van Deemter equation explains the various factors leading to band broadening.

$$\textbf{HETP} = \textbf{A} + \textbf{(B/u)} + \textbf{Cu}$$

where   A - Eddy diffusion
        B - Longitudinal diffusion
        C - Mass transfer
        u - Average mobile phase velocity

Eddy diffusion – (A term) – is caused by turbulence in the solute flow path and is unaffected by flow rate (velocity of mobile phase).

Longitudinal diffusion – (B term) – is the movement of the analyte molecules outwards from the centre to the edges of its band.

Increased flow rate can control B term, and peak broadening can be avoided.

Mass transfer – (C term) – is the movement of the analyte or mass transfer between mobile and stationary phases.

Increased flow rate increases the resistance to mass transfer leading to band broadening.

A consensus has to be arrived between these three terms to get the optimum flow rate with minimum HETP so that the efficiency of the column can be increased with good resolution of the peaks.

The van Deemter plot below indicates the desired plate height (H or HETP) and flow rate as optimum efficiency and flow rate, respectively.

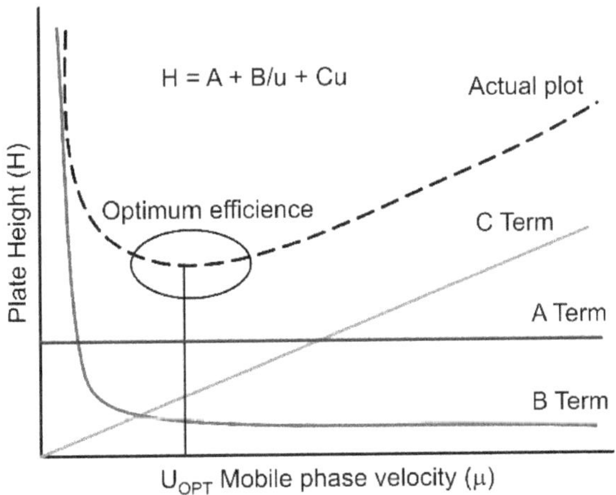

Fig. 7.3 Van Deemter plot.

Another equation given by the rate theory is $\textbf{H} = \sigma^2/\textbf{L}$ where σ - Standard deviation of band

H - HETP and is equal to H/dp,

Where dp is the particle size

$$H = (\sigma^2/L) \times dp$$

This equation gives the relationship between stationary phase particle size and efficiency of the column (HETP)

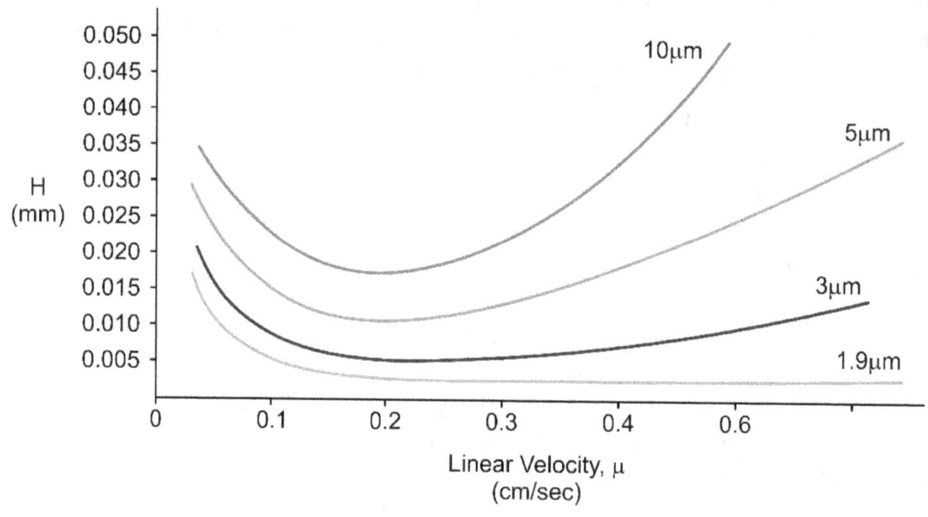

Fig. 7.4 Effect of particle size on Plate Height.

From the figure, it can be understood that **decreasing the particle size of the stationary phase yield better overall efficiency across a wide range of flow rates.**

We will discuss about different chromatographic techniques in detail in the forthcoming chapters.

CHAPTER 7A

Adsorption Column Chromatography

The process of separation of a mixture into individual components is called chromatography. Column chromatography was developed by American chemist D.T Day in 1900, and M.S.Tsweet, a Polish botanist in 1906, used it for the separation of plant pigments. It is the most useful and common method for separating solids and liquids. In adsorption column chromatography, the stationary phase is solid, and the mobile phase is a liquid.

PRINCIPLE

Adsorption is the principle behind adsorption column chromatography. Adsorption is a surface phenomenon where a solute adheres to the surface of another material, forming a thin layer. This layer is often only one molecule thick. The solute molecule that gets adsorbed on the surface is known as adsorbate, and the surface on which adsorption occurs is known as adsorbent. The separation of the compounds depends on the interaction between adsorbents and adsorbates. The stationary phase is the adsorbent and the sample to be separated is the adsorbate.

The stationary phase used is packed in a column, and hence the name column chromatography. The principle behind column chromatography is **adsorption**, and the components are separated based on **differential adsorption**. The mixture of samples to be separated is dissolved in the mobile phase and introduced into the column. The rate at which the components move along the column depends on their relative affinities for the stationary or mobile phases. More affinity for the stationary phase makes the components to move slow, and the components with more affinity for the mobile phase move fast and gets eluted rapidly. Because of this difference in the affinities, the components are separated. This kind of adsorption interaction between the stationary phase and the solute is reversible in nature.

COMPONENTS OF COLUMN CHROMATOGRAPHY

1. Stationary phase
2. Mobile phase
3. Column

1. **Stationary phase:** The adsorbent is used as a stationary phase. The requirements of the adsorbent are
   1. The particle should have a uniform particle size. 60-200μ
   2. The particle should be spherical in shape.
   3. It should not be soluble in the solvent and mobile phases used.
   4. It should be colourless.
   5. It should be able to separate a wide variety of compounds.
   6. It should be inexpensive and readily available.

   Selection of stationary phase

   The proper stationary phase is selected based on the factors like
   (i)   Nature of the compounds to be separated.
         E.g., the polar components can be separated on silica, and non-polar compounds are separated using alumina.
   (ii) Length of the column: For shorter columns, a strong adsorbent is used. For longer columns, a weak adsorbent is used.
   (iii) Difference in affinities between components: For components with similar affinities, a stronger adsorbent is used.
         The weaker adsorbent is used when there is a large difference in affinities of components.
   (iv) Adsorbent quantity: Adsorbate: adsorbent ratio - 1:20 or 1:30

   Types of adsorbents: The important adsorbents used in column chromatography are silica, alumina, calcium carbonate, calcium phosphate, magnesia and starch, etc

   Polar compounds are separated using silica, and alumina is used for non-polar compounds. The adsorbents can be classified as given in table 7A.1

Table 7A.1 Classification of adsorbents.

Weak	Medium	Strong
Sucrose	$CaCO_3$	Activated silica gel
Starch	Ca phosphate	Activated alumina
Inulin	Mg carbonate	Activated charcoal
Talc	Mg oxide	Activated magnesia
Sodium carbonate	Calcium hydroxide	Fuller's earth

2. **Mobile phase:** The mobile phase in the column is very important because it performs several functions. The functions of the mobile phase are:
   (i)   as a solvent- introduces the sample into the column.
   (ii) as a developing agent- to develop the bands in the column
   (iii) as an elute – to elute the components from the column.

   Selection of mobile phase

   (i)   The sample components must be soluble in the solvent system.

(ii) The mobile phase should have a low boiling point so that the sample can be recovered.

(iii) The solvent of appropriate polarity should be selected because the polarity of the solvents affects the relative rates of movement of the components.

The mobile phase is chosen based on these properties.

The common solvents used as mobile phases are given in the increasing order of polarity (elution strength): Petroleum ether, carbon tetrachloride, cyclohexane, Carbon disulphide, ether, acetone, benzene, toluene, ethyl acetate, chloroform, methanol, ethanol, water, pyridine, organic acids like acetic acid, etc.

The mobile phase can be a single solvent or a mixture of solvents.

3. **Column:** The column is usually made up of neutral glass since it should not be affected by solvents. It is used as a support for the stationary phase. Column dimensions are Length: Diameter ratio- 10:1; 30:1 or 100:1

The length of the column used depends on the following factors:

1. The affinity of the solute (adsorbate) towards the adsorbent used.
2. No. of components to be separated.
3. Nature of the adsorbent.
4. Sample quantity.

Longer columns are used for better separation.

Methodology of column chromatography

The different steps involved in column chromatography are

I. Preparation of column
II. Sample introduction
III. Elution
IV. Detection.

I. **Preparation of the column:** Before packing into the column, the adsorbent is activated by heating to increase the adsorption capacity. Heating removes the adsorbed water from the adsorbent. Activation of adsorbent is done by heating silica gel to 110°C and alumina to 400°C for one hour. To start with the preparation, a glass wool/ cotton or asbestos pad is inserted at the bottom of the clean glass column. Then the column is packed with adsorbents to get a uniform packing. A paper disc is placed on top of the packing to avoid the disturbance of the packing during the introduction of the sample or mobile phase. Disturbance in the adsorbent packing causes irregular band separation.

The packing of the column is done by two methods

1. Dry packing
2. Wet packing

1. **Dry packing:** In this technique, the dry adsorbent is filled in the column by tapping, and the mobile phase is added till the equilibrium is reached in the column.

   Demerits:
   1. Air bubbles are trapped in the column, and the column is tapped to remove air bubbles.

2. Cracks appear in the packed adsorbent.
3. Clear band separation may not be obtained.

2. **Wet packing:** This is the most commonly used ideal technique. The adsorbent is made as a slurry by mixing with the solvent and pouring it into the column in portions. Slurry should be added carefully to avoid cracks and get a uniform adsorbent setting without the air bubbles. The solid adsorbent settles at the bottom, and the solvent covers the adsorbent layer on top.

Advantages:
1. The packing is uniform and suitable for separation.
2. The resolution of the bands is good.

II. **Sample introduction:** The sample mixture is dissolved in a small volume of the mobile phase or suitable solvent. The entire sample is added at once into the column, and the sample gets adsorbed onto the top of the column to form a zone. The individual components are separated from this zone by elution.

III. **Elution (Developmental technique):** The recovery of the components is carried out by a process called elution. The development of the chromatograph zone of the separated components from the column is called elution. The solvent used is called eluent. The components are the eluates.

There are two elution techniques:
1. Isocratic elution
2. Gradient elution

1. **Isocratic Elution:** The same solvent or same solvent composition is used throughout the chromatographic separation.

   E.g., chloroform throughout the separation, Pet ether: Benzene in the ratio 1:1 throughout separation

2. **Gradient elution:** In gradient elution, the composition of the eluting solvent can be varied gradually during the separation according to the nature of the components being separated. Solvents of increasing polarity/elution strength are used during the separation. A low polar solvent is used initially, and the polarity of the eluent is increased gradually by using more polar solvents.

   e.g., elution starting with benzene, chloroform, ethylacetate, and then changing to methanol.

   Developmental techniques other than elution are **frontal analysis** and **displacement analysis**.

   In frontal analysis, the original solution containing the sample is used for displacement. The sample is fed continuously into the chromatographic bed, and no additional mobile phase is used. The sample components displace each other in the order of least affinity for the chromatographic medium.

   In displacement analysis, another solvent that is more strongly adsorbed than the original solvent is used. It is a procedure in which the mobile phase contains a more strongly retained compound (the displacer) than the components of the sample. The sample is fed as a finite slug.

Chromatograms are plots of the concentration of the components in the eluents against time.

IV. Detection: The coloured components can be observed visually, and they can be collected individually and detected. If the isolated components are colourless, anyone of the following detectors can detect the compounds. Sometimes developers like hydrogen sulphide and potassium ferrocyanide are used to visualize the separation as bands.

1. Optical detectors
   (a) UV- visible detector – If the sample absorbs UV/ visible light
   (b) Fluorescence detectors: For the samples which are emitting radiations.
2. Refraction index detector
   It detects the components based on the RI difference between the mobile phase with compound and without compounds.
3. Flame ionization detector/ argon detector
4. The other simple technique is collecting the fractions, evaporating the solvent and weighing the residue.
5. Conducting detectors used for ionized substances.
6. Thin-layer chromatography is used to monitor the fractions.

Factors affecting column chromatography

1. **Column dimensions:** Column efficiency can be improved by increasing the length: diameter ratio. E.g., 20:1 or 30:1 ratio can be increased to 100:1
2. **Adsorbent particle size:** Adsorbent activity depends on the surface area of adsorbent. Surface area increases with the decrease of particle size.
   Thus by decreasing particle size, adsorbent activity can be increased.
3. **Column temperature:** Elution speed increases with an increase in temperature, but the adsorbent activity decreases. Hence a compromise should be made between the elution and adsorbent activity. Room temperature is used for all samples.
4. **Column packing:** Uniform column packing leads to better separation of components.
5. **Solvent Nature:** More viscous solvents have less flow rate. Viscosity is inversely proportional to flow rate. Solvents of low viscosities give better separation.
6. **Pressure:** By increasing pressure above the column, the retention time of the analyte can be decreased.
7. **Adsorbent activity:** Adsorbent should be activated by heating to remove water.

Methodology of column chromatography

A tubular column similar to a burette is used. The column is packed with adsorbents like finely divided silica, alumina, charcoal, and calcium carbonate. The mixture to be separated is applied to the top of the column. Then the mobile phase is allowed to pass through the column under the influence of gravity. Equilibrium is established between the solute, adsorbent and the mobile phase. Different components have different interactions with the stationary and mobile phases.

They move at different rates according to their affinity towards the stationary phase along the column, and separation is achieved. The least adsorbed component moves faster and gets eluted faster. Most adsorbed components move slower and get eluted later. The eluting solvent may be the same as the mobile phase or different from the mobile phase. The elution order of organic compounds with the increasing polarity of the solvent is given as

Alkanes < Alkenes < Ethers < halogenated hydrocarbons < Aromatic hydrocarbons < aldehydes and ketones < esters < alcohols < amines < carboxylic acids.

Normal phase chromatography (NP)

In the normal phase, the stationary phase is polar in nature, and the mobile phase is non-polar. Based on the 'like dissolves like' principle, the more polar component interacts more strongly with the stationary phase, moving slowly. The non-polar compound moves rapidly and gets eluted faster.

Reversed-phase column chromatography (RP)

In RP column chromatography, the stationary phase is non-polar, and the mobile phase is polar. Here, the non-polar compound moves slowly because of the increased interaction with the stationary phase. Polar compounds get eluted faster in the polar mobile phases (like dissolves like). Column chromatography mostly uses RP type of separation.

APPLICATIONS

(i) Column chromatography is used for the separation of
1. Inorganic ions (cation and anions)
2. Organic molecules (dyes, diastereoisomers)
3. Tautomers
4. Racemic mixtures
5. Geometrical isomers
6. Plant extracts
7. Drugs
8. Pharmaceuticals

(ii) Used in the purification process to remove impurities

(iii) Isolation of active constituents from plant extracts.

(iv) Estimation of drugs in formulations

CHAPTER 7B

Partition Column Chromatography

Partition chromatography is also known as Liquid-Liquid chromatography. If the mobile phase is gas, it is called Gas-Liquid chromatography. Partition chromatography is the process of separation in which the components of the mixture gets distributed between two immiscible liquid phases based on their partition coefficients in a chromatographic column.

The stationary phase is a thin layer of liquid that is adsorbed on the surface of a substrate. The mobile phase liquid is immiscible with stationary phase liquid. The solute equilibrates between the mobile phase and stationary phase liquids.

THEORY

Partition coefficient (*P*) is the ratio of concentrations of a solute in a mixture of two immiscible solvents at equilibrium. This ratio compares the solubilities of the solute in these two liquids. The partition coefficient generally refers to the concentration ratio of un-ionized species of the solute. Most commonly, one of the solvents is hydrophilic(polar), and the second solvent is hydrophobic(non-polar). Hence the partition coefficient measures the nature of the chemical substance, whether it is hydrophilic (water-loving) or hydrophobic (water-fearing).

In partition chromatography, the separation of the components of a mixture depends on the **partition coefficient of the individual** components for the given set of solvents. Thus the basis of separation in partition chromatography is the partitioning of the individual components of the mixture to be separated in the given set of immiscible solvents.

The column is packed with finely divided inert solid particles, and each particle is coated with a thin layer of a liquid. This liquid acts as the stationary phase. The mobile phase is a liquid that is immiscible with the stationary phase liquid. The mixture to be separated is dissolved in the mobile phase and poured on the top of the column. Depending on their partition coefficient, the components will dissolve either in the stationary phase or in the mobile phase. The partition coefficient K is given by the formula

$$K = \frac{\text{Concentration of solute in stationary phase}}{\text{Concentration of solute in mobile phase}}$$

Solute with a higher partition coefficient will be dissolved more in the stationary phase.

In partition chromatography, usually, water is held as a thin film on support as a stationary phase. The mobile phase is an organic solvent saturated with stationary phase liquid (water).If the partition coefficient of the components of the mixture for the solvents is different, then the solutes are separated. An equilibrium is established between the solutes and the solvents.

The partition coefficient $k_x = \dfrac{X_s}{X_m}$

where

X_s- concentration of solute in stationary phase

X_m- concentration of solute in the mobile phase

If K_x is high, the sample spends more time in the stationary phase and vice versa.

The capacity factor $K_x' = \dfrac{\text{Total moles of x in stationary phase}}{\text{Total moles of x in the mobile phase}}$

The volume of mobile phase required to elute 50% of the solute is called retention volume (V_r)

$$K_x' = \frac{V_r}{V_m} - 1$$

where

V_r - Retention volume.

V_m -Volume of mobile phase in the column

This equation is obtained by including the volume of the mobile phase and stationary phase in the calculation.

Retention time is the time taken for the solute to elute from the column. It is the time between the injection of the sample and the detection of the sample by the detector.

It can also be defined as the time taken for the solute peak to appear at the column exit or the time taken to elute 50% of the solute.

In terms of retention time, the equation can be written as,

$$K_x' = \frac{t_r - t_o}{t_o} = \frac{t_r - 1}{t_o}$$

where

t_r – retention time of sample

t_o – time for the solvent molecule or unretained solute to pass through the column

This is the fundamental equation applied in gas-liquid and liquid-liquid chromatography.

The equation shows the relationship between the retention time and the solubility of the sample in the stationary and mobile phases.

Separation in partition column chromatography is done based on the principle that the partition coefficient of the components is a constant for the given set of stationary and mobile phases.

Solid Supports

Commonly used solid supports are silica gel, diatomaceous earth like kieselguhr, celite and cellulose. The stationary phase used with these solid supports is water.

Selection of Mobile and stationary phase

The stationary phase and mobile phase should be immiscible with each other. The stationary and mobile phase combinations used with the solid supports are given in table 7B.1.

Table 7B.1 Combinations of stationary phase and mobile phases.

Supports	Stationary Phase	Mobile Phase	Applications
Celite	Water	$CHCl_3$ or CCl_4	C_1-C_4 alcohols
Silica gel	Water	$CHCl_3$ /butanol	C_2-C_8 fatty acid, acetylated amino acids
Silica gel	Water	Iso – octane	Phenols
Cellulose	Water	Acetone, HCl Methanol/butanol,$CHCl_3$	Inorganic ions, phenols
Starch	Water	Propanol or Butanol HCl	Amino acids

If the stationary phase is soluble in the mobile phase, there is loss of stationary phase with continued use. To avoid this, the mobile phase should be pre-saturated with the stationary phase.

Solvent systems

Water, aqueous buffers are used as the stationary phase, and organic solvents are used as the mobile phase. The factors to be considered in the selection of the solvent systems are

1. Complexation
2. pH
3. Mixed solvents
4. Non-equilibrium effects

Common solvent combinations used in partition chromatography are given in table 7B.2.

Table 7B.2 Solvent combinations used in partition chromatography.

Stationary phase	Mobile phase
Water	Alcohols(n-butanol, isobutanol)
A mixture of water and acid	Hydrocarbons (Benzene,toluene, cyclohexane, hexane)
A mixture of water and alkali	Chloroform
A mixture of water and buffers	Chlorosilanes and silicones
Aqueous alcohols (Methanol, Ethanol)	Ethyl acetate
Alcohols (Methanol, Ethanol)	Ethylene glycol, Monomethyl ether
Formamide	Methyl ethyl ketone
Glycols like ethylene, propylene and Glycerol	Pyridine

Advantages

1. Applicable for the separation of closely related compounds like homologous series.
2. It can be applied for low concentration of the solutes
3. Polar molecules can be separated.
4. Resolution is greater.

Disadvantages:

1. Miscibility problems of mobile and stationary phases.

Reverse-phase chromatography (RP)

In RP Chromatography, the stationary phase is non-polar, and the mobile phase is polar in nature, contrary to the normal phase chromatography. The most common stationary phase is silica gel over which organic material is bonded Eg:-C_{18} or C_8. RP partition chromatography gives efficient separation of more polar solutes.

APPLICATIONS

Partition column chromatography can be used in

1. Detergent removal from protein solutions
2. Separation of steroids, bile acids and micro toxins.
3. Removal of pesticides, Phenol and insecticides.
4. Determination of trace metal concentration from aqueous solution.
5. Resolution of amino acids after hydrolysis of protein.
6. Separation and analysis of closely related aliphatic alcohols.
7. Separation of sugar derivatives.
8. Separation of carboxylicacids.

Table 7B.3 Comparison between Adsorption and Partition Chromatography.

Adsorption chromatography	Partition chromatography
Stationary phase: Solid Mobile phase: Liquid	Stationary phase: Liquid Mobile phase: Liquid
No need for solid support	A solid support is needed
Simple and more reproducible	Miscibility problems occur sometimes
Versatile and can be adapted to large scale procedures	More applicable to low concentrations

CHAPTER 8

Thin Layer Chromatography

Thin layer chromatography (TLC) is a type of planar chromatography used for the separation and identification of organic compounds and also to test the purity of the compounds. The separation and identification of a mixture of components into individual components in TLC is carried out using finely divided adsorbent, solid/liquid spread over a glass plate and a liquid mobile phase. The history of Chromatography is described as given below

Michael Tswett is credited as the father of liquid chromatography. Tswett developed his ideas in the early 1900s. In 1938 Izmailov & Shraiber described the basic principle and used it to separate plant extracts. Consden, Gorden & Martin started using filter papers to separate amino acids in 1944. In 1950 Kirchner used impregnated glass plate coated with alumina to identify terpenes. Ergon Stahl introduced standard equipment in 1958 for preparing TLC plates with uniform thin layers of known thickness.

PRINCIPLE

The principle of separation in TLC is **adsorption**. Adsorption is a surface phenomenon where a solute adheres to the surface of another material, forming a thin layer. This layer is often only one molecule thick. The solute molecule that gets adsorbed on the surface is known as adsorbate, and the surface on which adsorption occurs is known as adsorbent. The separation of the compounds depends on the interaction between adsorbents and adsorbates. The stationary phase is the adsorbent and the sample to be separated is the adsorbate.

A thin layer (0.25 mm) of adsorbent is coated onto solid support made of materials like glass, plastic and aluminium. The mixture of compounds to be separated is spotted on the adsorbent layer. **The mobile phase** solvent **rises** through the adsorbent (**against gravity**) by capillary action. The movement of the components is based on their affinities towards the adsorbent. **The component with more affinity for the stationary phase travels slowly and the component with lesser affinity moves faster, leading to the separation of components.**

A binder like plaster of Paris is mixed with the adsorbent to facilitate the binding of the adsorbent with the support.

COMPONENTS AND METHODOLOGY OF TLC

1. **Stationary phase:** The selection of stationary phase in TLC depends on the nature of the components of the sample mixture. To obtain satisfactory separation efficiency, the factors to be considered are particle size, size distribution and morphology of the particles. The various adsorbents available are
   Silica gel H (without binder), **silica gel G** (silica gel+ calcium sulphate),**silica gel GF** (Silica gel +binder+ Fluorescent indicator),
   Alumina-without binder (Neutral, Basic, Acidic), **Alumina with the binder,**
   Cellulose with and without binder,
   Kieselguhr G, Diatomaceous earth+ binder and polyamide.

2. **Support for the adsorbent:** The support on which the adsorbent coated, is made up of **glass plates** of dimensions 20cm × 20cm, 20 × 10 cm or 20 × 5 cm. **Aluminium foil, plastic film or sheets** are also used as supporting plates for adsorbents like silica gel, alumina and cellulose. Ready to use TLC plates are available with adsorbent layer of varying thickness coated on aluminium foil. The thickness of these plates varies from 0.1 to 0.2mm.The glass TLC plates are prepared using a TLC spreader.

3. **Preparation and activation of TLC Plates:** The slurry of the adsorbent in water is made and using this slurry, the TLC plates are prepared. The different methods used for the preparation of TLC plates are
   (a) Pouring
   (b) Dipping
   (c) Spraying
   (d) Spreading

   (a) **Pouring:** The prepared slurry is poured onto the glass plate placed on a levelled surface. It is spread uniformly on the glass plate. The plates are allowed to be set, then dried in an oven.
   Disadvantage: Uniformity in thickness not obtained.

   (b) **Dipping:** This technique is used for small plates. Two plates are held back to back and dipped into the slurry of the adsorbent at the same time. The slurry is prepared in chloroform or in other volatile solvents. The plates are allowed to set, then dried
   Disadvantages:
   1. The thickness of the adsorbent layer cannot be controlled.
   2. Uniformity in thickness may not be there.
   3. Large quantity of slurry is required even for the preparation of few plates

   (c) **Spraying:** The slurry is diluted, and the suspension of the adsorbent(slurry) is sprayed over the glass plate using a sprayer.
   Disadvantage: Difficult to get a uniform layer.

   (d) **Spreading:** This is the most suitable technique to get a uniform thin layer of adsorbent on glass plates. The operation of the spreading technique with the spreader is shown below.

There are two types of spreaders.

(i) Moving spreader type: In this type, after pouring the slurry, the spreader is moved over the plates spreading the adsorbents.

(ii) Moving plate type: The plates are made to move under the adsorbent reservoir, so that they are coated with adsorbent.

In both methods, the plates are air-dried to avoid cracks developing on the surface of the adsorbent.

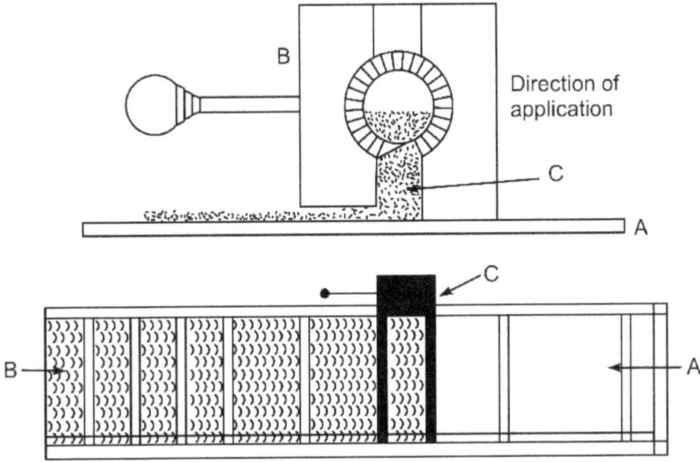

Fig. 8.1(a) Operation of thin layer spreader, **(b)** Tray alignment with partially coated plates A = Glass plate, B = Spreader, C = Slurry of the adsorbent.

Here, the slurry (C) is put in the spreader (B) and then moved along the direction of application onto the surface of the glass plate (A) to obtain a uniform layer.

In Fig (b), the slurry of the adsorbent is introduced into the aligning tray (C), which is then moved onto the glass plate (A) to obtain a uniform layer of adsorbent (B).

The glass plates are stacked on a base plate. The prepared slurry is poured into the reservoir of the TLC spreader. A knob is provided in the spreader to adjust the thickness of the adsorbent layer. The process is shown diagrammatically in Figures 8.1a and 8.1b.

For analytical purposes, 0.25mm thickness plate and 2mm thickness plate is used for preparative purposes.

After setting, the plates are activated by heating in an oven at 100°C to 120°C for one hour. Activation of the TLC plate is the removal of water/moisture from the surface of the adsorbent. This helps in retaining the adsorbent activity of the adsorbent.

4. Mobile phase: The choice of the mobile phase is dependent on

(a) Nature of the component to be separated

(b) Nature of the stationary phase

(c)　Chromatographic mode

(d)　Sample size

Mobile phases used are either pure solvents or a mixture of solvents. Solvents of increasing polarity are used.

An **elutropic series** ranks solvents by their relative abilities to displace solutes from a given adsorbent. The order of the elutropic series (increasing polarity) is given as

Pet.ether<Carbon tetra chloride<cyclohexane< carbon disulfide<Ether<acetone<Benzene Toluene<Ethyl acetate<Chloroform<Alcohol<water<Pyridine<organic acids<acid mixtures with pyridine or alcohol.

Usually the mobile phase composition is decided by trial and error method or sometimes from the literature.

5. **Sample application:** The sample solution should be of minimum concentration. Usually, 2-5μl of 1% solution of sample and standard are spotted on the plate using capillary for qualitative analysis and micropipette for quantitative analysis. Spots are placed 2cm above the base of the plate, and the spots should be as small as possible. When placing the plate inside the development chamber for elution, the spots should not be immersed in the mobile phase.

6. **Plate development:** For developing the plates, development chambers or glass beakers can be used. Nowadays, twin trough chambers are available, which are economical in using solvents (Fig 8.6). After pouring the mobile phase into the development tank, the tank is lined inside with a filter paper to saturate the atmosphere inside the tank with solvent vapours to avoid the **edge effect**. Reasons for the presaturation of the chamber is shown in fig 8.4a and figure 8.4b.

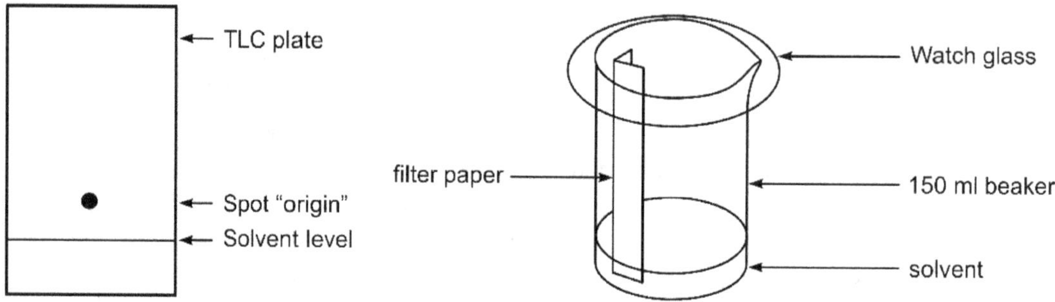

Fig. 8.2 Preparation & spotting of the TLC plate.　　**Fig. 8.3** Preparation of developing chamber.

The solvent front in the middle of the TLC plate moves faster than that of the solvent front in the edge as the mobile phase moves on the plate. This is called as **edge effect,** which makes the spots distorted. The Migration distance is more with the pre-saturated chamber than the non-saturated chamber. This is because, in a non-saturated chamber, the solvent migrating in the adsorbent layer vaporizes, leading to an edge effect. The edge effect is compared with normal TLC in figure 8.5.

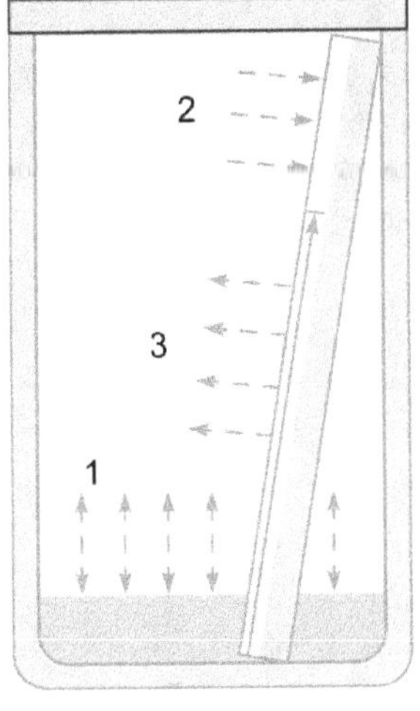

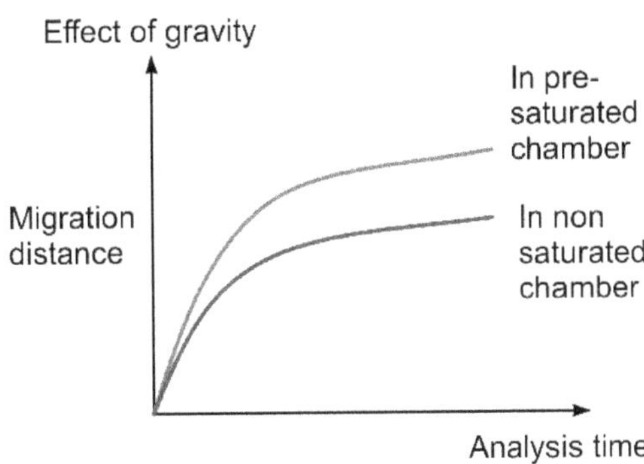

Fig. 8.4(a) Reasons for presaturation of chamber.

Fig. 8.4(b) Reasons for presaturation of chamber.

1. Solvent in Liquid-Vapour equilibrium
2. Solvent Vapour adsorbs on the layer
3. Solvent migrating from the layer vaporizes

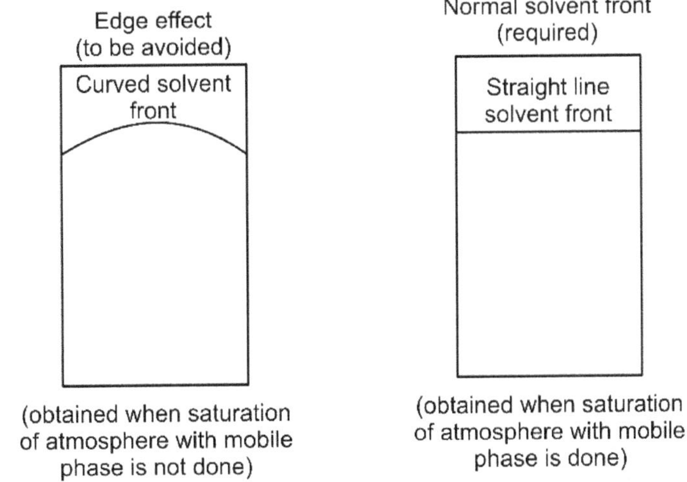

Fig. 8.5 Edge effect and normal TLC.

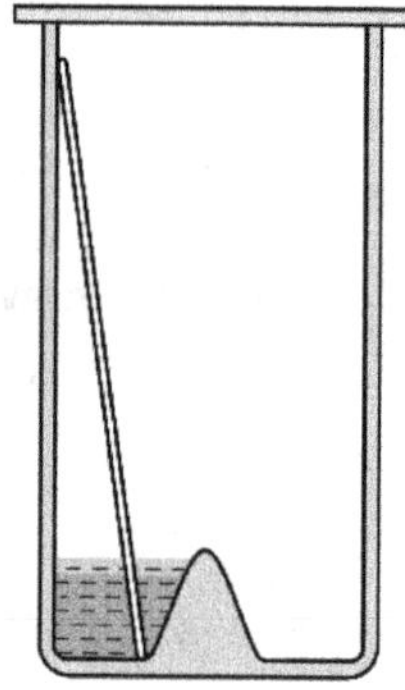

Fig. 8.6 Twin trough chamber.

After saturating the tank with solvent vapours, the prepared TLC Plate is placed in the development chamber and covered with a watch glass. The spots should be above the level of the mobile phase. It is kept aside undisturbed. The solvent rises up in the plate by capillary action. The solvent front is allowed to move up to $\frac{3}{4}$ th of the plate. The plate is removed from the tank, and the solvent front is marked with a pencil immediately.

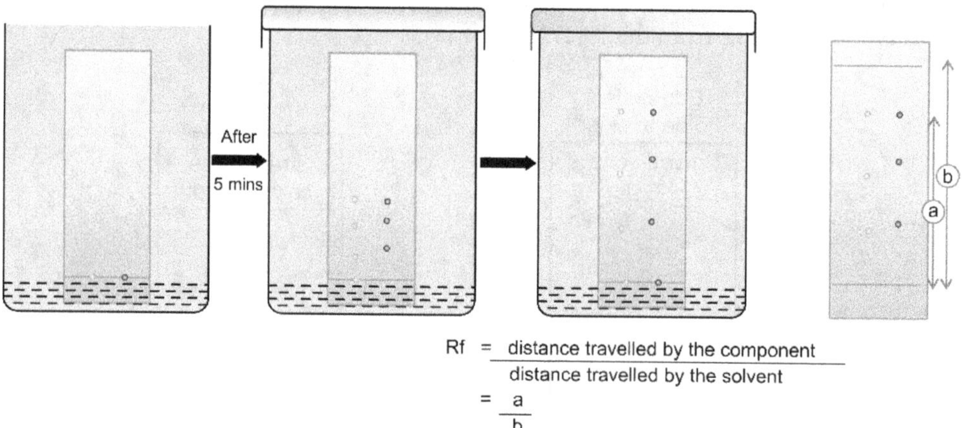

Fig. 8.7 Steps involved in TLC.

7. **Detection techniques:** After the development, the TLC plates are dried, and the analytes are detected. Detection of coloured samples can be done visually. The different types of detection techniques are
   1. Non-specific methods
   2. Specific methods
   1. **Non-specific methods:** These methods are used for the identification of compounds (qualitative analysis). The number of spots can be detected, but the nature of the compound cannot be detected.

(i) Iodine chamber method: Iodine crystals are added to the tank, and when the developed TLC plates are placed in the tank, the analyte spots appear as brown or amber-coloured spots.

(ii) Sulphuric acid reagent: The spraying reagent used is 70-80% of sulphuric acid with oxidizing agents like potassium dichromate, potassium permanganate or a few ml of nitric acid. After spraying this reagent, the plates are heated in an oven to give black spots, which are due to the charring of the compounds.

(iii) Using a fluorescent adsorbent: If the compounds are not fluorescent, they can be detected using a fluorescent adsorbent. The background is fluorescent and the compounds appear as dark spots when the plates are kept in the UV chamber. If the compounds are fluorescent in nature, they can be visualized as fluorescent spots in UV chamber, with a non-fluorescent adsorbent.

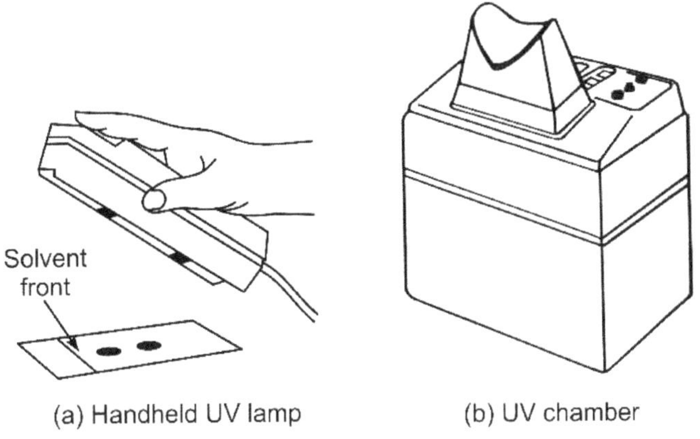

Fig. 8.8 UV detection methods.

2. Specific methods: Different detecting agents are used to identify the compounds. The agents used are specific for the given compound and are listed in table 8.1.

Table 8.1 Detecting agents with corresponding compounds.

Compounds	Reagent
Phenolic compounds and tannins	Ferric chloride
Amino acids	Ninhydrin in acetone
Alkaloids	Dragendroff's reagent
Cardiac glycosides	3,5-dinitrobenzoic acid
Aldehyde and ketone	2,4-dinitrophenyl hydrazine

The detecting techniques are further classified as

1. Destructive techniques: In these methods, the samples are destroyed by the reagent during detection. Hence sample recovery is not possible. E.g., Sulphuric acid reagent.

2. **Non-destructive techniques:** The sample is not destroyed during detection in these methods. Sample recovery is possible. E.g., UV chamber method, Iodine chamber method and densitometric method. In densitometry, the spot density is measured by a densitometer. This method is used for quantitative estimation. It is also called an *in-situ* method.

QUALITATIVE ANALYSIS

1. **R_f value:** To identify the compounds in TLC, a parameter called **R_f (retardation factor)** value is used. R_f value is the ratio of the distance travelled by the solute to the distance travelled by the solvent front.

$$R_f = \frac{\text{Distance travelled by the solute}}{\text{Distance travelled by the solvent front}}$$

R_f values range from 0 to1. R_f value is characteristic for each compound for the given set of stationary and mobile phases. The compound can also be identified by comparison with the R_f value of the standard.

The factors affecting R_f values and reproducibility are

1. Increase in **layer thickness** decreases R_f value, because the mobile phase moves slower up the plate.
2. Moisture on the TLC plate
3. Vessel saturation
4. Temperature
5. Depth of mobile phase
6. Nature of the TLC plate
7. Sample size
8. Solvent parameters.

An increase in these parameters causes an increase in R_f values typically.

2. **R_x value:** In case the solvent front runs off the plate, the ratio of the distance travelled by the sample to the distance travelled by the standard is measured, and it is called as **R_x value**

$$R_x = \frac{\text{Distance travelled by the solute}}{\text{Distance travelled by the standard}}$$

The value is always closer to 1.

3. **R_m value:** To find out the compounds of a homologous series, **ΔR_m values are determined.**

$$\Delta R_m = \log \frac{1}{R_f} - 1$$

In a homologous series ΔR_m values are constant.

QUANTITATIVE ANALYSIS

Direct method: The quantity of the components can be determined by the densitometric analysis of individual spots.

Indirect method: The individual spots are eluted with a suitable solvent and determined by methods like UV visible, fluorescence, flame photometry and electrochemical methods.

8. **Development techniques:** The actual separation takes place in the plate development of thin-layer chromatography. A solvent system is introduced to the sample on the TLC layer to separate the mixture into individual substances. Separation can be achieved through various methods. The most common TLC development techniques are explained below.

The different developmental techniques in TLC are

1. One dimensional development

   Vertical and Horizontal, and multiple developments

2. Two-dimensional development

1. **One dimensional development**

Vertical Development: This is the most common TLC development method. The plate is placed in a suitable TLC developing chamber such that the solvent wets the adsorbent layer below the starting line. The solvent rises up the layer due to capillary forces transporting the sample mixture. Once the solvent front has reached the predetermined height (10-15 cm for TLC and 3-7 cm for HPTLC), the plate is removed from the chamber. The solvent front is marked with a pencil, and the plate is dried.

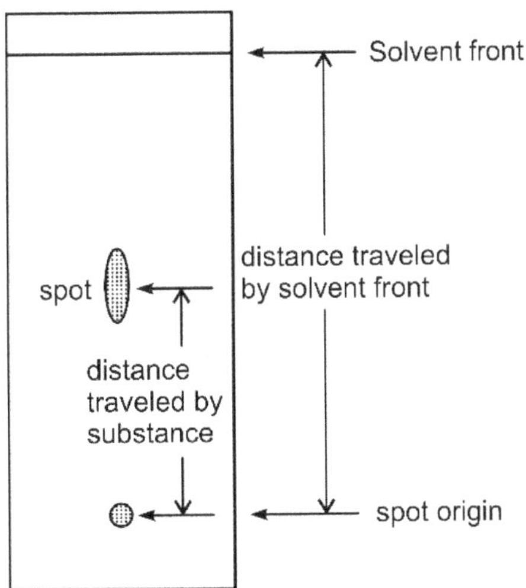

Fig. 8.9 Vertical development.

Horizontal Development: In this method, the plate is positioned horizontally inside the chamber, and the solvent is applied using a wick or capillary slit. Development can be performed on one or both sides of the TLC plate.

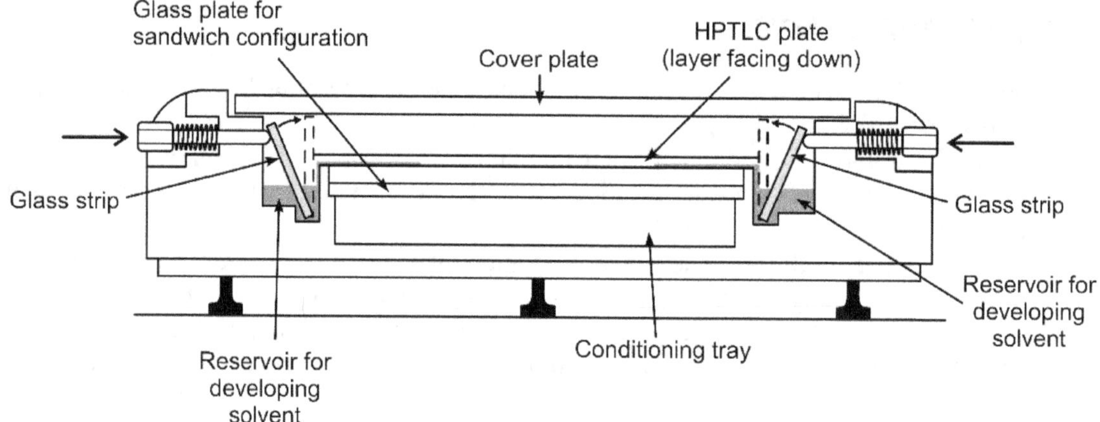

Fig. 8.10 Horizontal development.

Multiple Development: In this method, the TLC plate undergoes multiple developments with drying between each development. Re-concentration of the spots is caused by the repeated travelling of the solvent, producing elliptical shapes or narrow bands. The resolution is significantly improved for substances with R_f values below 0.5. Multiple developments can be performed over different separation distances, using the same solvent or different solvents of varying polarity. The different types are

Separate runs over the same migration distance

Stepwise, increasing

Stepwise, decreasing

Automated multiple developments, stepwise with a solvent gradient system

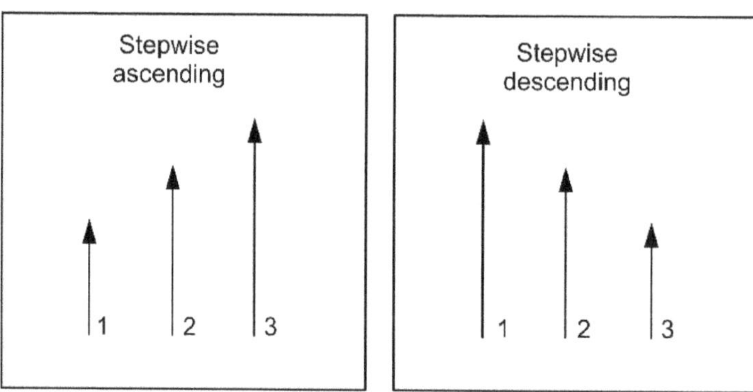

Fig. 8.11 Multiple developments.

Two-Dimensional Development
1. Two dimensions with one solvent system
2. Two dimensions with two solvent system

In two-dimensional TLC development, the sample is applied to a starting point in the corner of the TLC plate. The plate is placed in a normal chamber and developed from bottom to top. After drying, the plate is turned 90° and placed in another chamber with a different solvent and developed again. The end of chromatogram track from the first development is used as the starting line for the second development. The second development can be done with the same solvent or another solvent. Complex mixtures can be separated completely by two-dimensional chromatography.

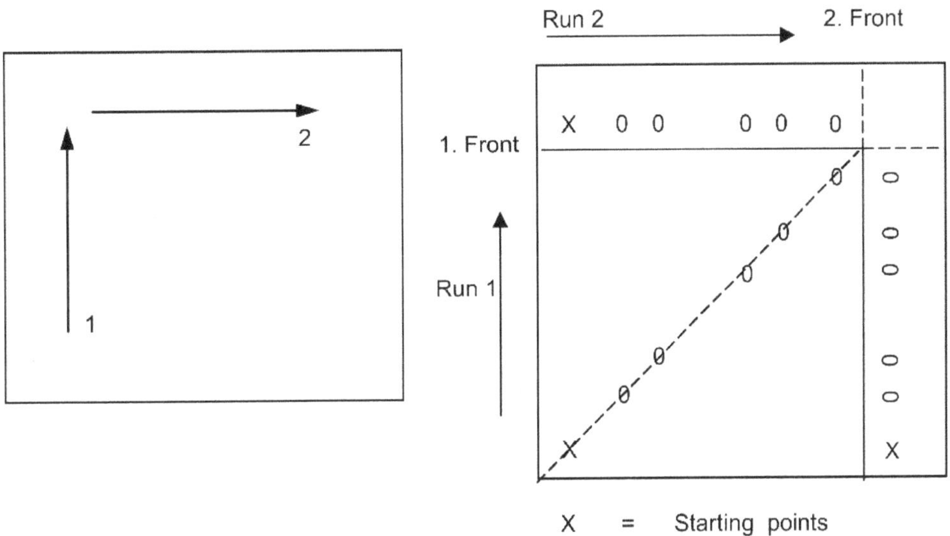

Fig. 8.12 Two-dimensional development.

APPLICATIONS OF TLC

1. Separation of a mixture of drugs of chemical, biological origin and plant extracts.
2. To check the purity of the samples and identification of drugs.
3. To purify the compounds
4. To check the completion of reactions
5. For identifying organic compounds. E.g., Acids, alcohols. Glycols, alkaloids, amines, proteins, peptides and antibiotics
6. Most widely used for identification tests, tests for purity and identification of related substances.
7. Separation of inorganic ions (cations and anions)
8. Separation of amino acids and vitamins
9. To detect decomposition products in drugs.
10. Quantitative analysis by TLC

    It is done by two methods

1. Analysis of fractions on the plate - Photo densitometry.
2. Determination of fractions after elution from the supporting material.

After elution, the samples are quantified using methods like UV-Visible, Fluorimetry etc.

Advantages
1. Simple equipment and low cost
2. Rapid technique
3. Wide choice of stationary phases
4. Easy recovery of separated components
5. High efficiency of separation
6. Easy detection
7. More sensitivity
8. Both preparative and analytical scales are possible.
9. Allows the usage of corrosive spray reagents
10. Economical usage of solvents and stationary phases.

Disadvantages
1. The stationary phase needs activation.
2. Cannot differentiate **isomeric and enantiomeric forms of a compound**.
3. A large quantity of slurry is required to prepare a few TLC plates.
4. TLC is applicable only to nonvolatile compounds, thus limiting its use.

HIGH-PERFORMANCE THIN-LAYER CHROMATOGRAPHY

This is a sophisticated and automated form of TLC used to separate, identify, and quantify a mixture of components. It is also used as a micro preparation technique. The principle of separation is adsorption, as in TLC. In HPTLC, the particle size of the adsorbent is small and the layer thickness is thin, giving better resolution than TLC.

A comparison between TLC and HPTLC techniques is presented in Table 8.2.

Table 8.2 Comparison between TLC and HPTLC techniques.

Parameters	HPTLC	TLC
Adsorbent layer thickness	100μm	250μm
Mean Particle size	5-6	10-12
Plate height	12 μm	30 μm
Migration distance	3-6 cm	10-15 cm
Separation time	3-20 min	20-200 min
Number of samples per plate	< 36	< 10
Sample volume	0.1-0.5 μl	1-5μl
Mobile phase volume	Less consumption	More
Sampling	Autosampler	Manual
Scanning and detection	UV, Visible, Fluorimetric detection is possible for both qualitative and quantitative analysis. The densitometric scanner is used for the detection	Detection techniques are applicable. But scanners are not available for densitometric analysis.

CHAPTER 9

Paper Chromatography

Paper chromatography is a separation technique in which the mixture of components is separated on a chromatographic paper using a liquid mobile phase. The two types of paper chromatography are

1. Paper adsorption chromatography
2. Paper partition chromatography

In paper adsorption chromatography, the stationary phase paper is impregnated with silica or alumina, and the solvent is the mobile phase.

In paper **partition** chromatography, the **moisture** which is present in the **pores of the cellulose fibres** of the filter paper acts as the **stationary phase** and an immiscible **solvent** is used as the **mobile phase**.

Paper chromatography generally refers to paper partition chromatography because most of the separations are carried out by this method.

PRINCIPLE

The principle of separation in paper chromatography is partition. The moisture present in the cellulose layers of the filter paper acts as the stationary phase, and organic solvents/ buffers are used as mobile phases.

A small drop of the sample solution is applied as a small spot on the filter paper, and it is dried. The filter paper is kept inside the chamber containing the mobile phase, which rises by capillary action, and as it reaches the spots of the sample, the mobile phase carries the components along with it. The rate of movement of the components is dependent on their partition coefficients. When the mobile phase rises to ¾ th of the paper, the paper is removed from the chamber, dried and detected using suitable reagents. The movement of the components is expressed in terms of migration parameters like the R_f value.

MIGRATION PARAMETERS

The position of the migrated components is indicated by the terms such as R_f, R_x and R_m.

1. **R_f value:** To identify the compounds in paper chromatography, a parameter called **R_f (retardation factor)** value is used. R_f value is the ratio of the distance travelled by the solute to the distance travelled by the solvent front.

$$R_f = \frac{\text{Distance travelled by the solute}}{\text{Distance travelled by the solvent front}}$$

R_f values range from 0 to 1. R_f value is characteristic for each compound for the given set of stationary and mobile phases. The compound can also be identified by comparison with the R_f value of the standard.

The factors affecting R_f values and reproducibility are

1. Nature of mobile phase.
2. Vessel saturation
3. Temperature
4. Sample size and nature of the sample
5. Nature of filter paper

An increase in these parameters normally causes an increase in R_f values.

2. **R_x value:** In case the solvent front runs off the plate, the ratio of the distance travelled by the sample to the distance travelled by the standard is measured, and it is called as **R_x value**

$$R_x = \frac{\text{Distance travelled by the solute}}{\text{Distance travelled by the standard}}$$

The value is always closer to 1.

3. **R_m value:** To find out the compounds of a homologous series, **ΔR_m values are determined.**

$$\Delta R_m = \log \frac{1}{R_f} - 1$$

In a homologous series ΔR_m values are constant.

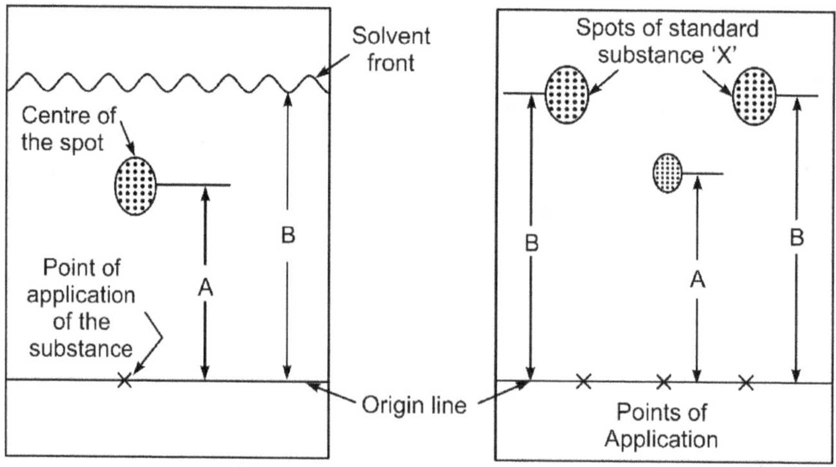

Fig. 9.1 Diagrammatic representation of R_f value. **Fig. 9.2** Diagrammatic representation of R_x value.

EXPERIMENTAL DETAILS FOR QUALITATIVE ANALYSIS

CHOICE OF PAPER USED (STATIONARY PHASE)

The choice of the filter paper depends on

(a) Nature of the sample to be separated (hydrophilic, hydrophobic or neutral).

(b) Nature of separation, whether qualitative or quantitative

(c) Analytical or preparative separation

The different types of Whatman filter papers are grades like No.1, No.2, No.3, No.4, No.20, No.40 and No.42 etc. These papers have different sizes, shapes, porosities and thicknesses.

TYPES OF PAPERS

A. **Pure cellulose papers (Whatman papers)**

B. **Modified cellulose papers**

   1. Chemically modified papers

     (a) Carboxyl papers (b) Acetylated papers

   2. Impregnated papers

   3. Loaded papers

     (a) Papers loaded with adsorbent powder

     (b) Papers loaded with Ion-Exchange celluloses

     (c) Papers loaded with synthetic organic Ion-Exchange resins

     (d) Papers loaded with inorganic Ion-Exchangers

C. **Glass fibretype papers**

  A. Pure cellulose papers (Whatman papers): These types of papers are prepared from cotton linters low in organic and inorganic impurities and uniform in physical characteristics. Whatman papers have α Cellulose 98-99%and 1% of mineral content

  B. Modified cellulose papers: Cellulose paper can be modified in several ways to alter its behaviour.

    1. Chemically modified papers (Hydrophobic papers)

    (a) Carboxyl papers: The exchange capacity of the paper is increased by increasing the carboxyl content (1.4%) by partial oxidation.

    They are used for efficient separation of polar substances, Cationic separation of potential amines and amino acids.

    (b) Acetylated papers: These papers are partially acetylated by treating with glacial acetic acid containing 5% acetic anhydride.

    They are used in Reverse phase chromatographic separation of lipophilic substances like steroids, insecticides, pigments and also metal cations.

    (c) Hydrophilic papers: e.g., Papers modified with methanol, formamide, glycol, glycerol etc.

2. **Impregnated papers:** Paper sheets are impregnated with oils to separate hydrophobic substances such as amines, lipids, steroids, vitamins, and pesticides (reverse phased chromatography)

3. **Loaded papers:** In these papers, powder, fibre, or resin are dispersed in a normal cellulose paper

    (a) Paper loaded with adsorbent powder, E.g. Silica papers, alumina papers, zirconia papers, kieselguhr papers.

    These papers are used to separate less polar and nonpolar substances (such as lipids, sterols, steryl esters, steroids, terpenes, dyes, insecticides, chloroplast pigments and inorganic ions).

    (b) Papers loaded with Ion-Exchange celluloses

    Chromatographic papers are coated with anionic or cationic exchangers

    (c) Papers loaded with synthetic organic Ion-Exchange resins:

    These papers are coated with micro pulverized ion-exchange resin, which is incorporated into α-cellulose pulp (reeve angle papers)

    (d) Papers loaded with inorganic Ion exchangers: e.g. zirconium phosphate, hydrous oxide, and ammonium molybdophosphate

C. **Glass Fiber Type Papers:** These papers are manufactured from fine borosilicate glass fibres. Detection reagents that are too corrosive for cellulose-based papers (e.g. chromic-sulphuric acid charring solution) can be used on these papers. Glass fibre type papers can be used in extreme conditions of temperature and acidity where cellulose papers are unstable.

MOBILE PHASES

Pure solvents, buffer solutions or a mixture of solvents can be used as a mobile phase. The mobile phases can be classified as Hydrophilic and Hydrophobic mobile phases.

Hydrophilic mobile phases

Iso propanol: ammonia: water	9:1:2
n-butanol: glacial acetic acid: water	4:1:5
Methanol: water	3:1 or 4:1
t-butanol: water: formic acid	40:20:5

Hydrophobic mobile phase

Kerosene: 70% Isopropanol,

Dimethyl ether: cyclohexane

Single, two or three-phase solvent systems can be used.

APPLICATION OF SAMPLES

The sample is dissolved in the mobile phase and applied using a capillary tube or micropipettes. Sample in µg quantities like 10-20µl are used for spotting (low concentration). Spots are applied on the origin line, which is drawn by a pencil,5cm from the bottom of the filter paper. Each sample is spotted 2cm away from each other.

DEVELOPMENTAL TECHNIQUES

The different developmental techniques used in paper chromatography are

  (i)   Ascending development
 (ii)   Descending development
(iii)   Ascending- descending development
 (iv)   Circular or radial paper chromatography
  (v)   Two-dimensional development

  (i)   **Ascending development:** It is otherwise called vertical development. The spotted (at the bottom) paper is kept vertically in the chamber containing the mobile phase. The solvent rises on the paper by capillary action against gravity.

 (ii)   **Descending development:** In this type of development, a special chamber holds the solvent at the top. The paper is spotted at the top, and the solvent is allowed to flow down the paper assisted by gravity. Hence the development is faster.

(iii)   **Ascending – Descending development:** It is a hybrid of both ascending and descending techniques. First, the ascending technique is allowed to proceed. The filter paper is folded over a glass rod, allowing the ascending technique to be converted to descending. In this technique, the distance of travel of solute components increase leading to better separation.

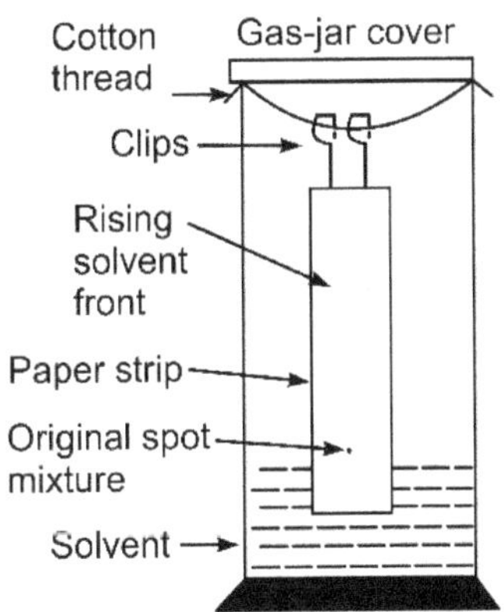

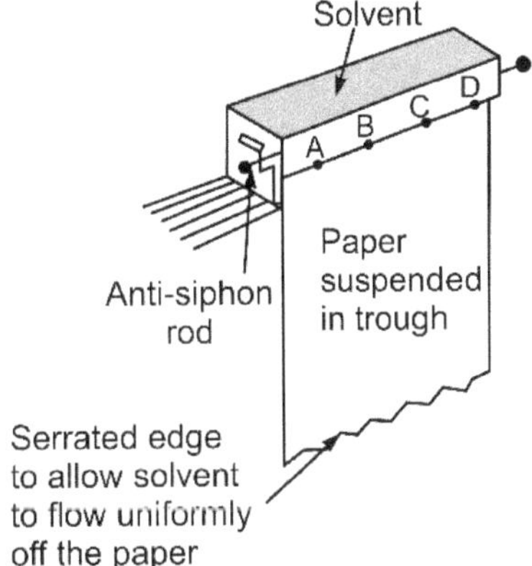

Fig. 9.3 Ascending development.   **Fig. 9.4** Descending development.

(iv) Circular / radial paper chromatography: In this technique, the development takes place radially. A circular filter paper is used. The sample is spotted in the centre. After drying, the paper is placed on a petri dish with the mobile phase, and a wick from the filter paper is dipped into the mobile phase. As the solvent rises through the wick and moves through the paper, it uniformly carries the components in all directions.

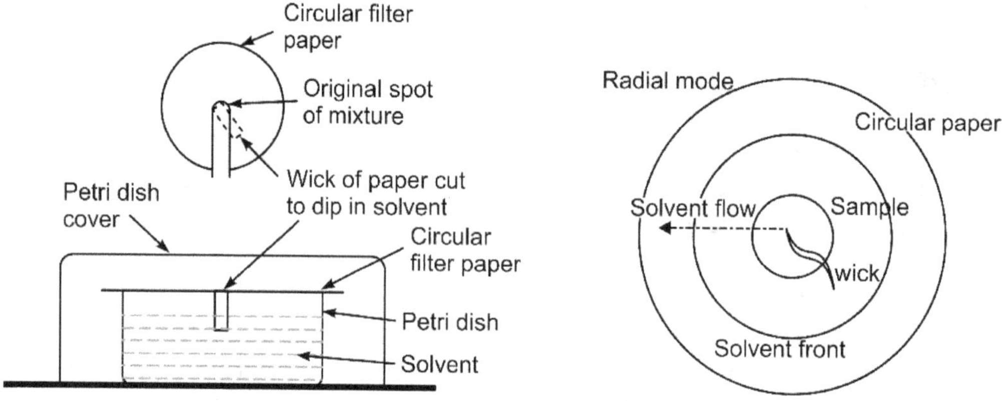

Fig. 9.5 Circular paper chromatography.

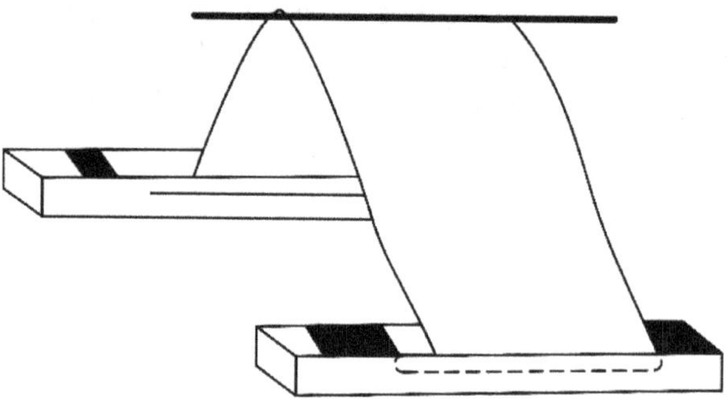

Fig. 9.6 Ascending – Descending development.

(v) Two-dimensional development

(a) Two dimensions with one solvent system

(b) Two dimensions with two solvent system

In two-dimensional paper development, the sample is applied to a starting point in the corner of the paper. The paper is placed in a normal chamber and developed once from bottom to top. After drying, the paper is turned 90° and placed in another chamber with a different solvent and developed again. The chromatogram track from the first development is used as the starting line for the second development. The second development can be done with the same solvent or another solvent. Complex mixtures can be separated completely by two-dimensional chromatography.

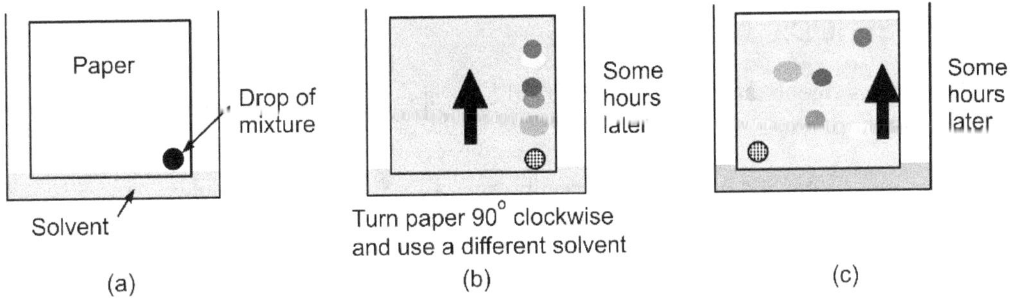

Fig. 9.7 Two-dimensional paper chromatography.

DETECTION TECHNIQUES

After the development, the chromatograms are dried and then detected. Detection of coloured samples can be done visually. The different types of detection techniques are

1. Non-specific methods
2. Specific methods
1. **Non-specific methods:** These methods are used for the identification of compounds (qualitative analysis).

    The number of spots can be detected, but the nature of the compound cannot be detected.

   (i) **Iodine chamber method:** Iodine crystals are added to the tank, and when the developed paper is placed in the tank, the analyte spots appear as brown or amber-coloured spots.

   (ii) **Using UV or Fluorescence:** If the compounds absorb in the UV region they can be detected using UV chamber. E.g., Antibiotics. Fluoresent substances can be detected by fluorimetric methods. E.g., Quinine.

    If the compounds are not fluorescent, they can be detected using a fluorescent adsorbent. The background is fluorescent and the compounds appear as dark spots when the plates are kept in UV chamber.

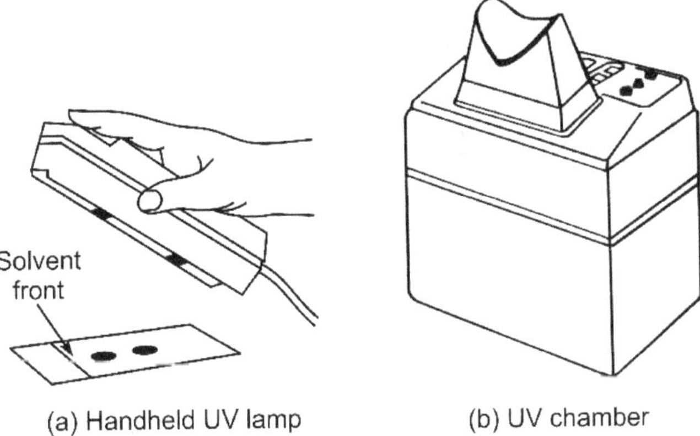

Fig. 9.8 UV chamber.

SPECIFIC METHODS

Different detecting agents are used to identify the compounds. The agents used are specific for the given compound.

Table 9.1 Some of the detecting agents for specific compounds.

Compounds	Reagent
Phenolic compounds and tannins	Ferric chloride
Amino acids	Ninhydrin in acetone
Alkaloids	Dragendroff's reagent
Cardiac glycosides	3,5-dinitrobenzoic acid
Aldehyde and ketone	2,4-dinitrophenyl hydrazine

The detecting techniques are further classified as

1. **Destructive techniques:** In these methods, during detection, the samples are destroyed by the reagent. Hence sample recovery is not possible. E.g., Sulphuric acid reagent.
2. **Non-destructive techniques:** The sample is not destroyed during detection in these methods. Sample recovery is possible E.g., UV chamber method, Iodine chamber method.

QUALITATIVE ANALYSIS

1. **R_f value:** To identify the compounds in TLC, a parameter called **R_f (retardation factor)** value is used. R_f value is the ratio of the distance travelled by the solute to the distance travelled by the solvent front.

$$R_f = \frac{\text{Distance travelled by the solute}}{\text{Distance travelled by the solvent front}}$$

   R_f values range from 0 to 1. R_f value is characteristic for each compound for the given set of stationary and mobile phases. The compound can also be identified by comparison with the R_f value of the standard.

2. **R_x value:** In case the solvent front runs off the plate, the ratio of the distance travelled by the sample to the distance travelled by the standard is measured, and it is called as R_x value

$$R_x = \frac{\text{Distance travelled by the solute}}{\text{Distance travelled by the standard}}$$

   The value is always closer to 1.

3. **R_m value:** To find out the compounds of a homologous series, **ΔR_m values are determined.**

$$\Delta R_m = \log \frac{1}{R_f} - 1$$

In a homologous series ΔR_m values are constant.

QUANTITATIVE ANALYSIS

The individual spots are eluted with a suitable solvent and determined by methods like UV visible, fluorescence, flame photometry and electrochemical methods.

Advantages of Paper chromatography

1. A simple and rapid technique
2. Inexpensive
3. Excellent resolving power.

Disadvantages of Paper chromatography

1. Only a minimum quantity of sample should be applied to avoid diffusion through paper leading to poor separation
2. Improper placement of paper leads to error in identification
3. If chamber saturation is not done, it will affect the resolution

Applications of Paper chromatography

1. Separation of organic and biochemical productss
2. Separation of drugs and natural products
3. Separation of carbohydrates, vitamins, antibiotics, proteins, alkaloids, glycosides, and aminoacids.
4. Identification of drugs and related products.
5. Identification of decomposition products and drug metabolites in blood, urine etc.

CHAPTER 10

Electrophoresis

Electrophoresis is a physical method of analysis in which the compounds that acquire charge in the presence of an electric field are separated. Electrophoresis is defined as the **movement of charged particles in an electric field**, leading to their separation. It is a separation technique that is based on the differences in the mobility of charged particles in an electric field. The technique was pioneered in 1937 by the Swedish chemist Arne Tiselius to separate proteins. It is one of the widely used techniques in molecular biochemistry, microbiology and biochemical research. It is an efficient technique for the separation of proteins in a mixture.

In practice, a positive (anode) and negative (cathode) electrode are placed in a solution containing ions. When a voltage is applied across the electrodes, solutes of different charges, *i.e.*, anions (negative) and cations (positive) move through the solution towards the electrode of opposite charge.

Fig. 10.1 Electrophoresis Apparatus.

PRINCIPLE

The basic principle is that the rate of migration (separation) of ions depends upon the e/m (charge to mass) ratio. Molecules having different molecular weights will have different charge/mass ratios. This fact is responsible for the differential migration of charged particles.

When a voltage is applied across the electrodes, a potential gradient (E) develops and is given as

$$E = V/d$$

where V – Voltage and d- the distance between electrodes.

The charged molecules migrate towards oppositely charged electrodes under the influence of an electric field. The positively charged molecules move towards the cathode and the negatively charged molecules move towards the anode.

The migration of molecules under the influence of current, experiences a frictional force. If the charge on the molecule is Q coulombs, then the velocity v of the charged molecule is given as

$$v = EQ/f$$

where

f is the frictional force which retards the mobility of the molecule.

The frictional force is dependent on size, shape, hydrodynamic volume of the molecule, matrix pore size and viscosity of the medium.

The **electrophoretic mobility, μ** of an ion is defined as the ratio of velocity to the potential gradient

$$\mu = v/E = Q/f$$

where

$$f = 6\pi r\eta \text{ (Stokes law)}$$

i.e.
$$\mu = Q/6\pi r\eta$$

where

Q – Charge on the molecule (esu)

μ - Electrophoretic mobility

r – Radius of the molecule (cm)

η - Viscosity of the medium (poise)

Thus in electrophoresis, molecules with identical or similar charges can also be separated if the molecules have different frictional forces.

Working Principle

Biological molecules exist in a solution as electrically charged particles at a given pH. They will exist as anionic (negatively charged/basic), Zwitterions and cationic (positively charged / acidic) ions. Zwitterions are amphoteric molecules having no charge on them.

The total charge of molecules depends on the pH. When electricity is applied to the medium containing biological molecules, they migrate differentially depending on their net charge & molecular size and thus, different proteins/DNA can be separated.

Depending on the kind of charge the molecules carry, they move towards the cathode or anode. An ampholyte becomes positively charged in acidic condition and migrate to the cathode. In alkaline conditions, they become negatively charged and migrate to anode. The electrophoresis mechanism is illustrated in figure 10.2.

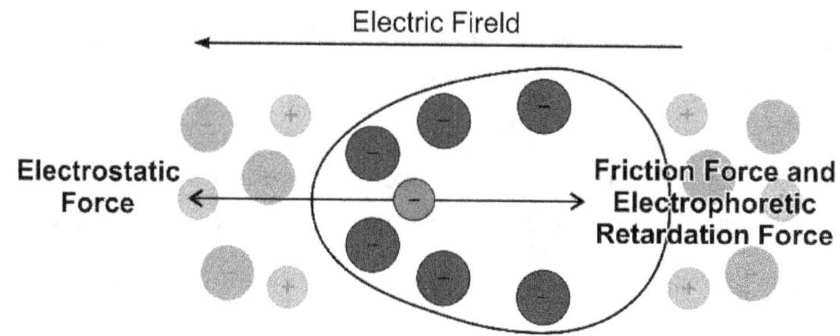

Fig. 10.2 Electrophoresis mechanism.

FACTORS AFFECTING ELECTROPHORETIC MOBILITY

1. Factors related to the electric field
2. Factors related to sample
3. Factors related to the buffer
4. Factors related to supporting medium

1. **Factors related to the electric field:**
   (a) **Voltage:** The rate of migration is proportional to VQ/d where V is Voltage, Q is the charge on the ion, and d is the distance between electrodes. Therefore, an increase in voltage causes a proportional increase in the rate of mobility.
   (b) **Current:** The migration distance of the ion is proportional to the current in the electrophoretic system.
   (c) **Resistance:** The current and rate of migration are inversely proportional to the resistance, which in turn depends on the nature of the medium, the buffer and its concentration. Resistance is directly proportional to the length of the supporting medium. It is inversely proportional to cross-sectional area and ion concentration of the buffer.
   (d) **Temperature:** Resistance decreases with an increase in temperature. An increase in temperature causes a decrease in viscosity of the liquid leading to decreased resistance and hence increased mobility. A decrease in resistance causes an increase in current and more heat will be produced. This heat evaporates the solvent and decreases the resistance.

2. **Factors related to the sample**
   (a) **Charge on the ion:** Migration rate increase with an increase in the charge on the ion

$$\mu = Q/ 6\pi r\eta$$

   Q – Charge on the molecule (esu)

   μ - Electrophoretic mobility

   E.g., Ca^{2+} will move faster than Na^+

   (b) **Size and shape of the molecule:** The size of the molecule is inversely related to migration. Smaller molecules exhibit faster migration rates when compared to larger molecules. Molecules with different shapes but similar sizes show different migration rates. The frictional resistance for the mobility depends on the shape of the molecule.

(c) **pK$_a$:** The mobility of molecules increases with a decrease in ionic strength. The degree of ionization is dependent on the pH of the buffer. Thus, the pH of the mobile phase is chosen based on the pK$_a$ of the analyte.

(d) **Diffusion:** Is based on Fick's law of diffusion, which states that the movement of ions occurs from the region of their higher concentration to the region of lower concentration

3. **Factors related to the buffer**

(a) **Composition of buffer:** The commonly used buffers are formate, acetate, citrate, borate, phosphate, EDTA and pyridine. The buffer should not interact with the sample and the support medium.

(b) **pH and ionic strength:** Mobility increases with decreased ionic strength. The pH of the buffer greatly affects the electrophoretic mobility, particularly when the sample is a weak acid/ weak base because the degree of ionization is dependent on pH.

4. **Factors related to the supporting medium:**

(a) **Adsorption:** Adsorption of solute molecules on the supporting medium causes the tailing of the sample. It decreases the migration rate and separation efficiency.

(b) **Electro osmosis:** It is caused by the relative charge produced between water molecules and the surface of the supporting medium. The charge may be caused by the interaction of the ionic molecules (sample) with the ionic side chain in the supporting medium.

E.g., sulphonate group in agar and carboxyl group on paper

5. **Viscosity of the medium:** The mobility of ions is inversely proportional to the viscosity of the medium.

TYPES OF ELECTROPHORESIS

Electrophoretic techniques can be broadly divided into two types

(i) **Zone electrophoresis**
   (a) Paper electrophoresis
   (b) Gel electrophoresis
   (c) Thin layer electrophoresis
   (d) Cellulose acetate electrophoresis

(ii) **Moving boundary electrophoresis**
   (a) Capillary electrophoresis
   (b) Iso-tachophoresis
   (c) Isoelectric focusing
   (d) Immunoelectrophoresis

MOVING BOUNDARY ELECTROPHORESIS

This method allows the movement of charged particles in a free moving solution in the absence of a supporting medium.

Disadvantages:

1. Complete separation is not possible.
2. Only the fastest and slowest components can be separated in pure form.
3. Preparative and quantitative analysis cannot be performed
4. Stabilisation of ion boundaries is complicated, and there is a need for specialized equipment

ZONE ELECTROPHORESIS

The disadvantages of moving boundary electrophoresis are eliminated by the use of a stabilizing medium in Zone electrophoresis. It involves the migration of charged particles which are supported on a relatively inert and homogenous solid or gel framework. Separated components are distributed into different zones in the support media.

E.g. of supporting media are paper, agar, cellulose, starch, gels, polyacrylamide gels

PAPER ELECTROPHORESIS

It is one of the simplest electrophoretic techniques which makes use of special filter papers as the supporting medium. This technique is mostly used for the separation of proteins and amino acids. It is a simple and economical method. Paper electrophoresis is adapted well to routine laboratory use.

PRINCIPLE

The principle is the same for all electrophoretic techniques as given above. The charge carried by the molecule depends on the pH of the medium. The mobility of the ion depends on the voltage applied and the charge on the ion.

WORKING PRINCIPLE

In this technique, Whatman filter paper No.1 or No.3 in a 3 or 5 cm wide strip is used as the supporting medium. The filter paper is moistened with the buffer, and the ends of the paper strip are immersed in beakers of electrolytes. The sample is applied as spots in the centre of the paper. A potential of 5V/cm of paper length is applied. The separation is carried out for several hours. After separation, the paper is dried and stained to get coloured bands.

INSTRUMENTATION

The instrumentation of electrophoresis consists of two parts.

1. Power pack
2. Electrophoretic cell

1. **Power pack:** The power pack provides DC and can control both voltage (0-500 V) and current output (0-150 amp).
2. **Electrophoretic cell:** The electrophoretic cell consists of
   1. Electrodes
   2. Buffer reservoirs
   3. Support for paper
   4. Transparent insulating cover

TYPES OF PAPER ELECTROPHORESIS

1. **Vertical paper electrophoresis** – Filter paper is mounted vertically. The sample is applied in the centre, and movement of the separated sample occurs in both directions depending on the charge.

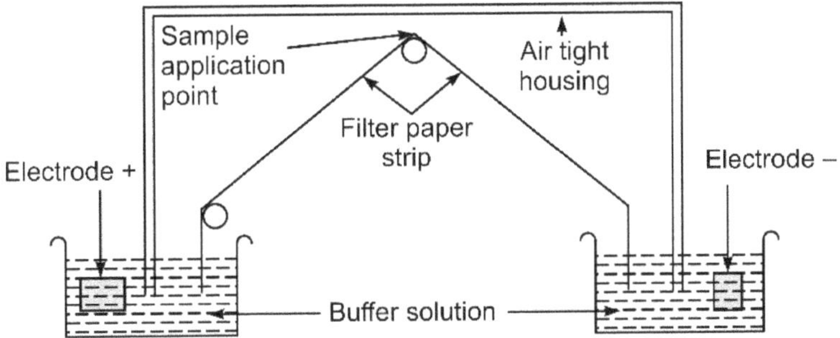

Fig. 10.3 Vertical Electrophoresis.

2. **Horizontal paper electrophoresis-** Filter paper is mounted horizontally. The sample is applied in the centre, and movement of the separated sample occurs depending on the charge.

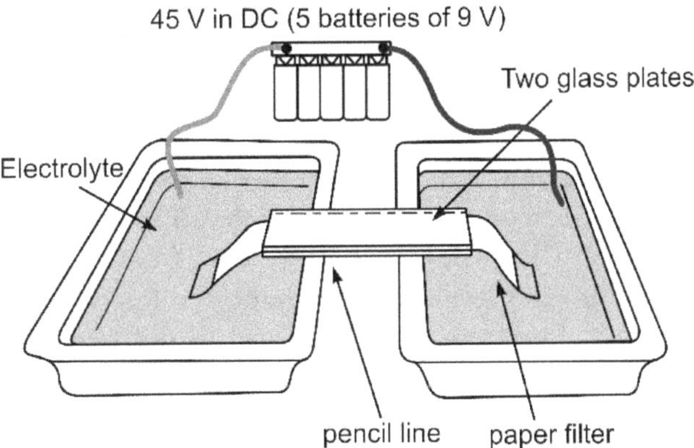

Fig. 10.4 Horizontal Electrophoresis.

3. **Continuous paper electrophoresis** – Predetermined sample volume is applied continuously on the centre of the paper. The suitable voltage causes migration and separation of samples.

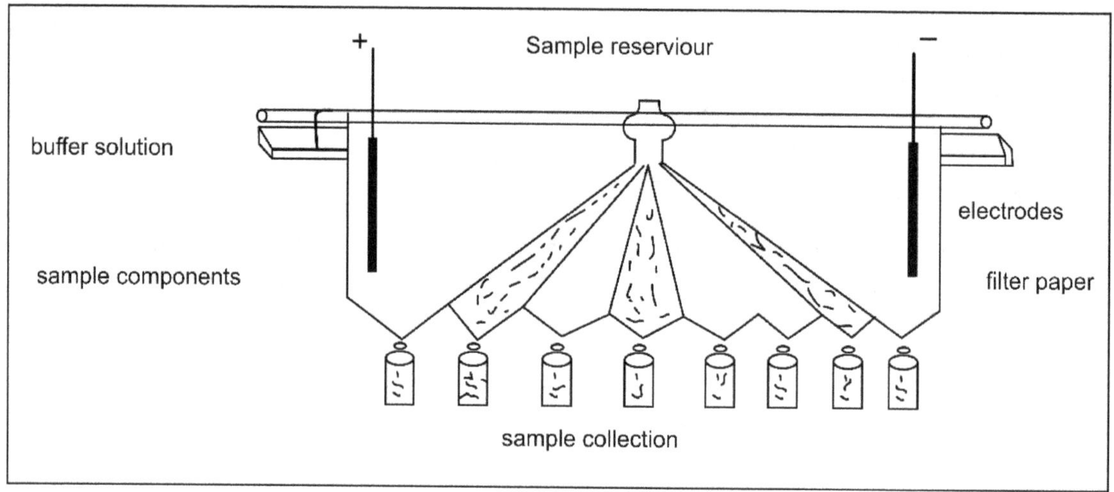

Fig. 10.5 Continuous paper electrophoresis.

Components

(i) **Paper:** Filter paper of good quality contains 95% α-cellulose. It has less adsorption capacity, and Grade No 1 or No 3 paper is used. The filter paper is moistened with buffer and both ends are dipped in buffer solutions in the electrode chambers.

(ii) **Electrodes:** Platinum electrodes are used. DC volt of about 8-15 V is applied

Based on the voltage applied, the electrophoresis can be classified as

Low voltage electrophoresis: Provides a voltage gradient of 5V/cm and a current of 0-150mA. Used for the separation of amino acids, proteins and sugars.

High voltage electrophoresis: Provides voltage gradient up to 100 V/cm. Rapid analysis consumes less time and bands are sharp. Useful for the separation of small ions derived from small peptides and amino acids.

(iii) **Buffers**: Buffers of different pH and ionic strength are used in separation. The buffer (pH) selection depends on the nature of the substance to be separated. The buffer is kept in the electrode chamber on both sides, and the ends of the filter paper are dipped in the buffer after moistening the paper with buffer.

The ionic strength of the buffer affects migration velocity. Mobility decreases with increasing ionic strength. Low ionic strengths lead to diffused bands. Usually, 0.05-0.5 ionic strength is used in separation.

E.g., Veronica buffer (baritone buffer) pH 8.6

Tris-acetate buffer –pH 7.6

Citrate buffer pH 3 or 6.8

Detection: The graph obtained for the unknown sample is compared with the graph of the standard sample for qualitative analysis.

The different techniques used for the detection are

1. **Fluorescence:** The staining reagents used are
   1. Ethidium bromide – used for the detection of DNA and RNA. It gives fluorescence under UV light.
   2. Fluorescamine – used to detect amino acids, peptides and proteins.
   3. Dansyl chloride - used in the detection of amino acids, peptides and proteins
2. **UV absorption:** Detection is carried out between 260-280nm. Used for the detection of amino acids, peptides and proteins.
3. **Staining**
   1. Proteins are stained with Dansyl chloride, Fluorescamine, Bromophenol blue in acetic acid.
   2. Nucleic acids are stained with Ethidium bromide
   3. For polysaccharides, iodine is used as the staining agent.
      After staining, different biomolecules appear as distinct bands within the gel.

Advantages:
1. Simple technique
2. Economical
3. More number of samples can be analysed simultaneously.

Disadvantages:
1. Time-consuming
2. Certain compounds like proteins, hydrophilic molecules cannot be resolved due to the adsorptive properties of paper leading to tailing and distortion of component bands.
3. Low field strength of 20V/ cm only can be used. It requires a very efficient cooling system.
4. Electro osmosis

Applications:
1. Serum analysis for diagnostic purposes.
2. Analysis of muscle proteins, egg white proteins, milk protein, snake and insect venom can be carried out
3. Alkaloids identification
4. Environmental studies.

GEL ELECTROPHORESIS

The supporting material used in this technique is a gel. The gel is polymerized in the shape of thin slabs and it has wells to load the sample.

Gel electrophoresis is carried out by two methods
1. Vertical gel electrophoresis
2. Horizontal gel electrophoresis

1. **Vertical gel electrophoresis:** Vertical gel electrophoresis contains a stacking gel and a resolving gel. The stacking gel concentrates proteins that are loaded into the well so that

the proteins can start to migrate simultaneously. After stacking, the resolution gel separates the proteins based on the molecular size. The top chamber of this system contains the cathode, while the bottom chamber contains the anode. The negatively charged linear strands of proteins migrate toward the anode (top to bottom) upon separation. Vertical gel electrophoresis is usually chosen for separating protein due to the ease of preparation of polyacrylamide gel vertically.

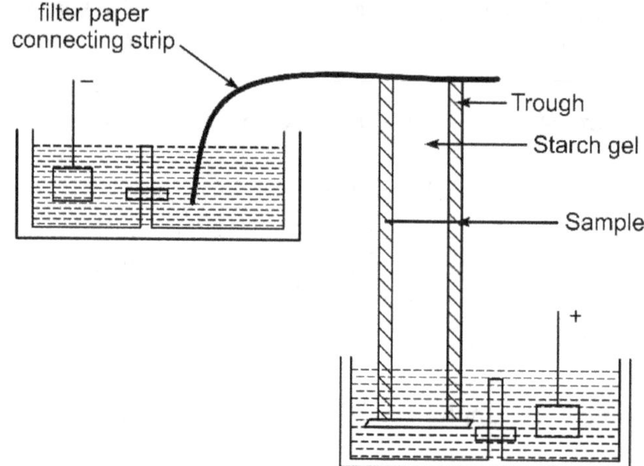

Fig. 10.6 Vertical Gel Electrophoresis.

2. **Horizontal gel electrophoresis:** Horizontal gel electrophoresis is used to separate DNA and RNA. Prior to separation, the DNA and RNA are loaded into the wells of agarose gel in the electrophoresis chamber. The chamber is connected to a cathode and anode, through which an electric field is created by an external power supply. DNA and RNA are negatively charged, they migrate from a negative end to a positive end.

The various types of the gel used in gel electrophoresis are

1. **Agarose gel electrophoresis**

2. **Polyacrylamide electrophoresis**

3. **Starch gel electrophoresis**

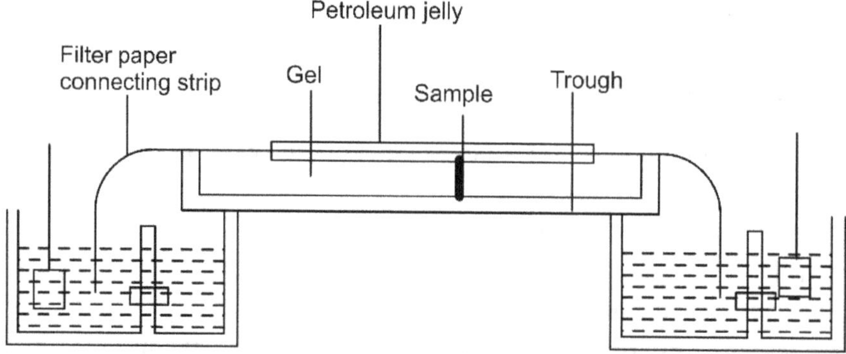

Fig. 10.7 Horizontal starch gel electrophoresis.

1. **Agarose gel electrophoresis:** Agarose gel is a polysaccharide obtained from seaweeds. It is a linear polymer composed of repeating units of D and L – galactose. Agarose melts at 90° and forms a gel at 40°C. Gel formation results in the formation of a mesh of channels with a diameter of 50-200nm. The agarose gel formation is shown in figure 10.8.

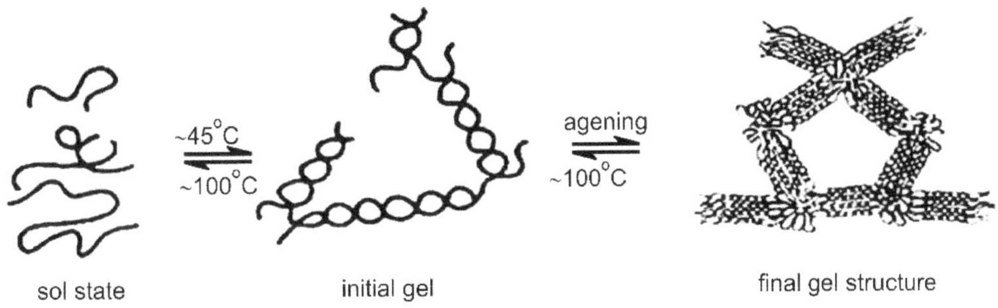

Fig. 10.8 Formation of Agarose gel.

Sample (DNA) is pipetted into the sample wells. The application of an electric current causes the negatively-charged DNA to migrate towards the cathode, positive end. The rate of migration is proportional to size. Small fragments move more quickly. DNA is visualized by including a dye in the gel. E.g., Ethidium bromide. DNA fragments take up the dye as they migrate through the gel. Examination under ultraviolet light causes the DNA to fluoresce.

The various steps involved in Agarose gel electrophoresis is shown in figure 10.10

Applications of Agarose Gel Electrophoresis

1. Estimation of the size of DNA molecules
2. Analysis of PCR products, e.g. in molecular genetic diagnosis or genetic fingerprinting
3. Separation of DNA and RNA.
4. Agarose gels are used for the purification of DNA fragments.

Advantages

1. Agarose gels are simple and rapid to prepare.
2. The gel is easily poured and does not denature the samples.
3. The samples can be recovered.

Disadvantages

1. Sometimes gels can melt during electrophoresis.
2. The buffer can become exhausted.
3. Occasional unpredictability in the running of DNA and RNA material

2. **Polyacrylamide gel electrophoresis:** Polyacrylamide gels are chemically cross linked gels formed by the polymerization of acrylamide with a cross linking agent, usually N, N'-methylenebisacrylamide. Polyacrylamide gel electrophoresis **(PAGE)** is a technique widely used in biochemistry, forensic chemistry, genetics, molecular biology and biotechnology to separate biological macromolecules, usually proteins or nucleic

acids according to their electrophoretic mobility. The process of ion migration in PAGE is shown in figure 10.9.

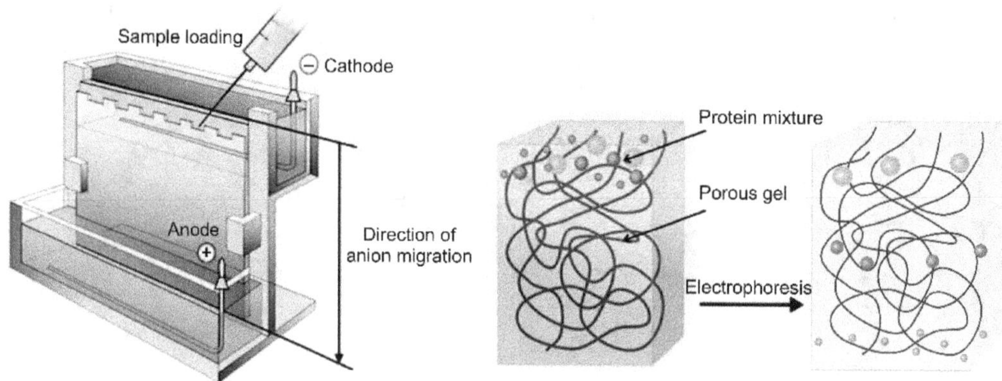

Fig. 10.9 Polyacrylamide gel electrophoresis (PAGE).

The most commonly used form of polyacrylamide gel electrophoresis is the Sodium dodecyl sulphate Polyacrylamide gel electrophoresis **(SDS- PAGE)**.It is used mostly for the separation of proteins.

The mobility of a substance in the gel depends on both charge and size. Hence the general electrophoresis techniques cannot be used to determine the molecular weight of biological molecules. To overcome this difficulty, different protein molecules with different shapes and sizes are denatured with SDS so that the proteins lose their secondary, tertiary or quaternary structure.

The proteins are covered by SDS and are negatively charged. When these denatured protein molecules are loaded onto the gel and placed in an electric field, they migrate towards the anode (positively charged electrode).They have acquired uniform charge by treatment with SDS, and hence the electrophoretic mobility depends primarily on size. The protein molecules are separated by a molecular sieving effect based on size. After staining (protein-specific), the size of a protein can be calculated by comparing its migration distance with that of a known molecular weight ladder (marker).

Buffer: Buffers allow the current to be carried through the sample, and they resist pH changes in the overall solution. The choice of the buffer depends upon the isoelectric point of the analyte. In DNA electrophoresis, buffers like TAE (Tris-acetate-EDTA) and TBE (Tris-borate-EDTA) are used. In protein electrophoresis, SDS (sodium dodecyl sulfate) is commonly used. These buffers facilitate the separation of the samples into readable gels.

Dye: A dye is used to monitor the migration of DNA into a gel or during gel electrophoresis. It is known as DNA gel loading dye. Eg: bromophenol blue, Ficoll 400, Xylene cyanol and glycerol. It is utilized as a colour indicator to monitor the migration of DNA in gel electrophoresis. The loading dye contains Ficoll or glycerol that gives density to the DNA sample. Henceforth, DNA cannot come out and diffuse in the buffer. DNA migration can be strictly monitored. The loading dye is the dye that is used for making the DNA markers, whereas the **tracking dye** is used to stain the DNA. The loading dye is used in the agarose and polyacrylamide gels, whereas the tracking dye is used in the agarose gel. The bromophenol blue is used as a loading dye, whereas the SYBR, an asymmetrical cyanine dye, is used as a tracking dye.

Applications:

1. Measuring molecular weight.
2. Peptide mapping and estimation of protein size.
3. Determination of protein subunits or aggregation structures.
4. Estimation of protein purity.
5. Analysis of the number and size of polypeptide subunits.

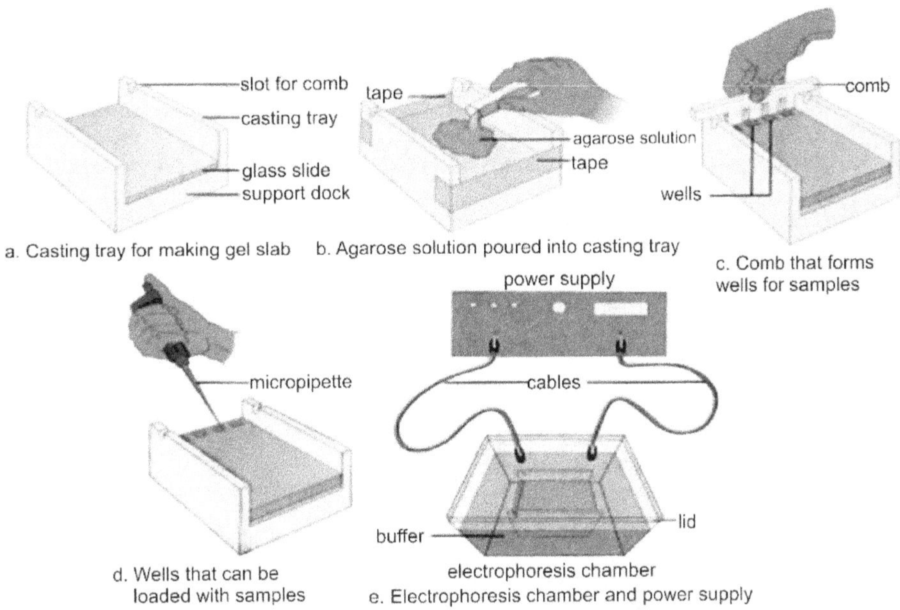

Fig. 10.10 Steps involved in Agarose gel electrophoresis.

Advantages of PAGE

1. Stable chemically cross linked gel with greater resolving power (Sharp bands)
2. Large quantities of DNA can be separated without significant loss in resolution
3. The purity of DNA recovered is extremely high.
4. The pore size of the polyacrylamide gels can be changed easily by changing the concentrations of the two monomers.
5. Low molecular weight fragments can be separated

Disadvantages of PAGE

1. Gels are tedious to prepare and handle.
2. Requires a longer time for preparation than agarose gels.
3. Monomers are toxic
4. Gels often leak, and a new gel is needed for each experiment

3. **Starch gel electrophoresis:** Starch gels are slightly more opaque than agarose or polyacrylamide types and are used for protein electrophoresis. They are formed from partially hydrolysed potato starch. A suspension of granular starch is boiled in a buffer to get a clear colloidal suspension. The semisolid gel is formed on cooling due to the intertwining of the branched chains of amylopectin. Petroleum jelly is used to avoid swelling and shrinking.

Advantages:

1. High resolving power and sharp zones are obtained.
2. Reasonable yield is obtained in the case of proteins
3. Can be used for analytical as well as preparative electrophoresis.

Disadvantages:

1. Electroosmotic effect.
2. Variation in pore size from batch to batch.

TWO-DIMENSIONAL GEL ELECTROPHORESIS (2-D ELECTROPHORESIS)

In the first dimension, proteins are resolved according to their isoelectric points (PI) using immobilized pH gradient electrophoresis (IPGE), isoelectric focusing (IEF), or non-equilibrium pH gradient electrophoresis. (Horizontal separation)

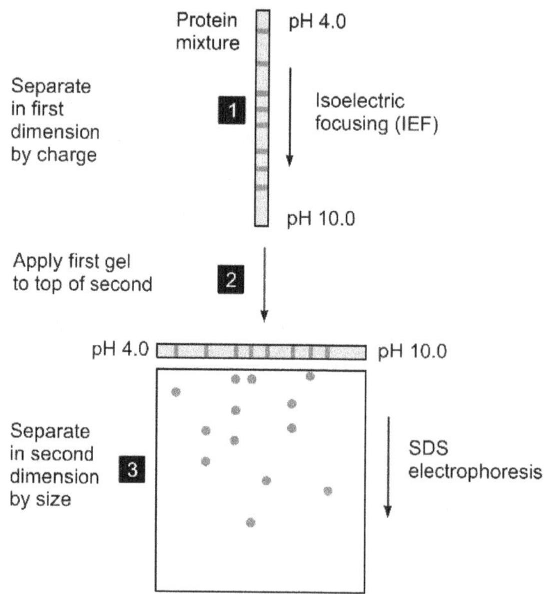

Fig. 10.11 Steps involved in 2D electrophoresis.

In the second dimension, proteins are separated according to their approximate molecular weight using SDS-PAGE. (Vertical separation).

CAPILLARY ELECTROPHORESIS

Capillary electrophoresis(CE) is an analytical technique used to separate ions based on their electrophoretic mobility under the applied voltage. The separation medium used is a capillary tube.

In capillary electrophoresis, the conducting buffer is filled in a capillary tube whose inner diameter is 25–75 μm. Samples are introduced in one end of the capillary tube. As the sample migrates through the capillary, the components separate and elute from the column at different rates. A component with a lower migration rate will be retained and have a longer retention time.

PRINCIPLE

The basic principle is that the rate of migration (separation) depends upon the e/m (charge to mass) ratio. Molecules having different molecular weights will have different charge/mass ratios. This fact is responsible for the differential migration of charged particles.

When a voltage is applied across the electrodes, a potential gradient (E) develops and is given as

$$E = V/d$$

where

V – Voltage and d- the distance between electrodes.

The charged molecules migrate towards oppositely charged electrodes under the influence of an electric field. The positively charged molecules move towards the cathode, and the negatively charged molecules move towards the anode.

The migration of molecules under the influence of current experiences a frictional force. If the charge on the molecule is Q coulombs, then the velocity v of the charged molecule is given as

$$v = EQ/f$$

where

f is the frictional force which retards the mobility of the molecule.

The frictional force is dependent on size, shape, hydrodynamic volume of the molecule, matrix pore size and viscosity of the medium.

The **electrophoretic mobility, μ** of an ion is defined as the ratio of velocity to the potential gradient

$$\mu = v/E = Q/f$$

where

$$f = 6\pi r\eta \text{ (Stokes law)}$$

i.e. $$\mu = Q/\, 6\pi r\eta$$

where

      Q – Charge on the molecule (esu)

      μ - Electrophoretic mobility

      r – Radius of the molecule (cm)

      η - Viscosity of the medium (poise)

Thus in electrophoresis, molecules with identical or similar charges can also be separated if the molecules have different frictional forces.

The capillary is filled with the buffer of specific pH for the separation of the analyte. The sample is introduced into the capillary by pressure injection or by electrokinetic injection. When a high voltage is generated (> 300V/cm), the sample components move through the capillary at different speeds. Positive components migrate to the negative electrode, and the negative components migrate towards the positive electrode and are detected by the detector.

WORKING

The size and charge on the ion determine its mobility in electrophoresis. The charge is dependent on pH. The pH can be changed to alter the migration of components to get the best separation. The pH is decided based on the pK_a values of the components.

Capillary is made up of fused silica-containing surface silanol groups (Si-O-H), which is slightly acidic in nature. In higher pH values, lots of negative charges are present on capillary walls (Si-O$^-$) due to the loss of proton (H$^+$). The capillary wall, which has a negative charge, develops a second layer of cations attracted to it. The inner cation layer is stationary, while the outer layer is free to move along the capillary. The applied electric field causes the free cations(+ve) to move towards the cathode(-ve), creating a powerful bulk flow. When the cations are travelling, they drag the buffer fluid with them. This is called **Electro Osmotic Flow** (EOF).

The EOF is flowing towards the cathode, negative electrode. During a separation, uncharged molecules move at the same velocity as the electroosmotic flow (with very little separation). Positively-charged ions move faster, and negatively-charged ions move slower.Thus +ve, -ve and neutral components all can be detected and separated. The diagrammatic representation of electroosmotic flow is given in figure 10.12

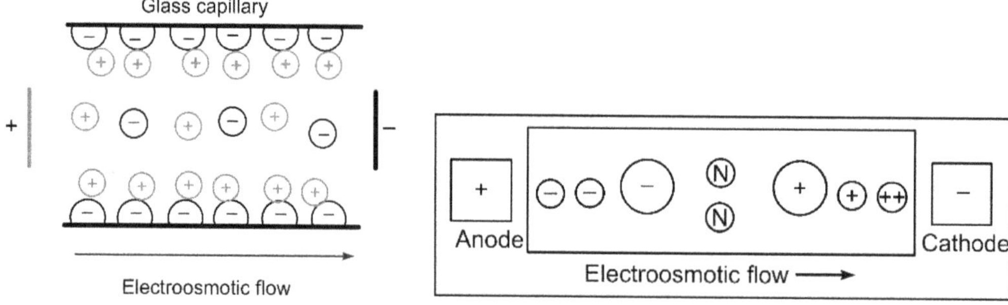

Fig. 10.12 Electroosmotic flow.

DIFFERENT MODES OF CAPILLARY ELECTROPHORESIS

The different modes of capillary electrophoresis are

1. Capillary zone electrophoresis (CZE)
2. Capillary gel electrophoresis (CGE)
3. Micellar electrokinetic capillary chromatography (MEKC)
4. Capillary electrochromatography (CEC)
5. Capillary isoelectric focusing (CIEF)
6. Capillary isotachophoresis (CITP).
7. Electro chromatography (EKC)

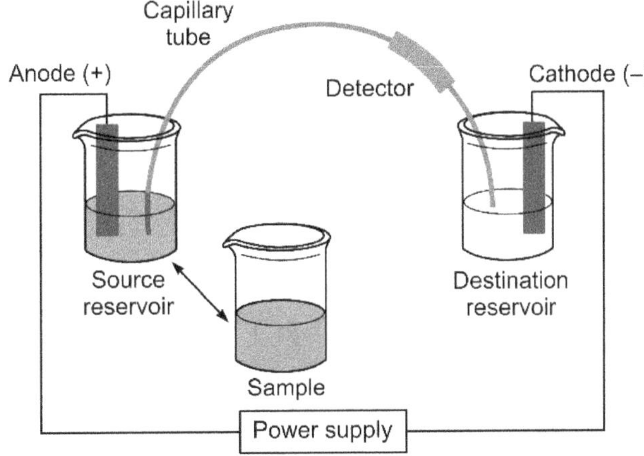

Fig. 10.13 Schematic diagram of instrumentation of Capillary electrophoresis.

The capillary electrophoresis can be classified into continuous and discontinuous systems, as shown in flow chart 10.14.

A continuous system has a background electrolyte acting throughout the capillary as a buffer. The continuous system is further classified into kinetic (constant electrolyte composition) and steady-state (varying electrolyte composition) processes.

A discontinuous system keeps the sample in distinct zones separated by two different electrolytes.

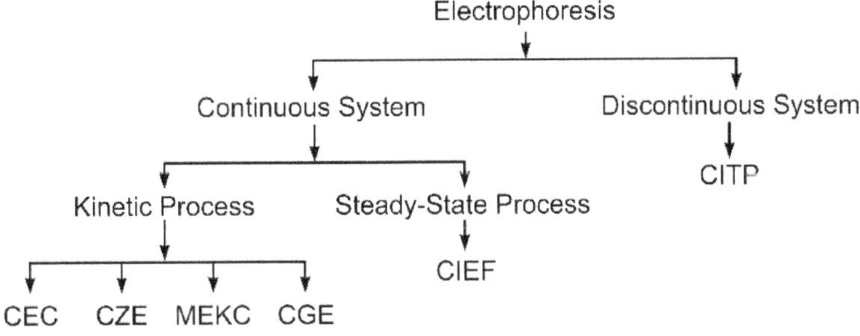

Fig. 10.14 Flow chart showing the types Capillary of electrophoresis.

1. **Capillary Zone Electrophoresis (CZE):** Capillary Zone Electrophoresis (CZE) is the most commonly used technique. It is also known as **free solution capillary electrophoresis**. The separation is based on the differences in electrophoretic mobility. The velocity of the ion is directly proportional to the electrophoretic mobility and the magnitude of the electric field. The electrophoretic mobility is directly proportional to the charge on the molecule and inversely proportional to the viscosity of the solvent and radius of the atom.

2. **Capillary Gel Electrophoresis (CGE):** A gel matrix is used in the capillary. Separation is based on the size of the components. Different sized but same mobility components are separated by this technique. CGE is used for protein and DNA separation. Smaller molecules move faster and the bigger molecules move slowly in CGE.

3. **Micellar electrokinetic capillary chromatography (MEKC):** Solutes are partitioning between micelles and the solvent in MEKC. When a surfactant is added to a solution, aggregates of surfactant molecules are formed above the critical micelle concentration. These aggregates are called micelles. The aggregates have polar negatively charged surfaces outside (Hydrophilic) and non-polar surfaces inside the micelle(hydrophobic).

   Hydrophobic molecules spend more time with micelles, while the hydrophilic molecule migrates faster with the solvent.

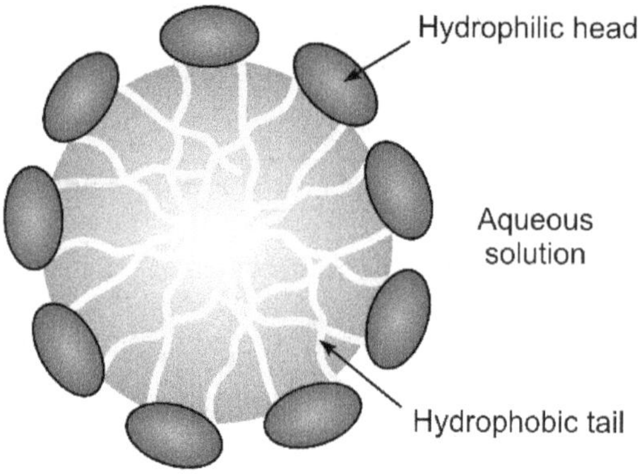

Fig. 10.15 Micelle formation.

Factors that affect the electroosmotic flow in MEKC are pH, surfactant concentration, additives, and polymer coatings of the capillary wall.

4. **Capillary electrochromatography (CEC):** The capillary is packed with silica particles. The separation principle is the same as column chromatography. Electroosmotic flow is present due to the charges on the silica. The flow obtained is a plug-type flow.

5. **Capillary isoelectric focusing (CIEF):** CIEF is commonly used to separate peptides and proteins. These molecules are zwitterionic in nature because they contain both positive and negative charges. Functional groups attached to the main chain and the pH of the environment determines the charge on the molecule. Each molecule has a specific isoelectric point (pI). When the surrounding pH equals this pI, the molecule carries no net charge (zwitterion form).

At pH lower than pI – molecule exists as a positive ion

pH higher than pI – molecule exists as a negative ion.

Since the charge on the molecule changes with pH, a pH gradient can be used to separate molecules in a mixture. A pH gradient is maintained across the capillary and a voltage is applied. The anode (+ve voltage) of the capillary is situated in an acidic solution (low pH), while the cathode (-ve voltage) is situated in a basic solution (high pH). The components of the mixture migrate at different rates to the pH value that equals the pI value of each component. At pI, the molecule exists as a zwitterion, and its migration is stopped.

Pressure is applied to move the complete pH gradient through the capillary, and the components pass through the detection window.

6. **Capillary isotachophoresis (CITP):** It is a discontinuous system based on the principle of moving boundary electrophoresis. The technique of isotachophoresis depends on the development of potential gradients.

   A leading electrolyte (e.g. chloride) with a higher mobility than the analytes and a trailing electrolyte (e.g. glycinate) with lower mobility is used. The solution in which the separation takes place is normally an aqueous medium containing sucrose to provide a higher density to the solution.

   The technique of isotachophoresis depends on the development of a potential gradient.

   The analytes are positioned between the electrolytes and when the voltage is applied, they migrate in order of decreasing mobility. This establishes the potential gradient and from that point on, all the analytes move at the same speed.

   Individual zones border one another but represent completely separated components without overlap.

   In isotachophoresis, no buffer is mixed with the sample, so current flow is carried only by charged sample ions.

7. **Electro chromatography (EKC):** It is used to separate chiral compounds by differential interaction of enantiomers with cyclodextrins.

   Advantages of CE
   1. High separation efficiency
   2. Rapid technique and high resolution of separation
   3. Detection is simple.
   4. Easy operation and low waste generation,

   Disadvantages
   1. The small diameter of the capillary causes dissipation of heat, leading to increased diffusion.
   2. Sample sticking to capillary walls
   3. Inconsistent retention times.

APPLICATIONS

1. **DNA fingerprinting:** Using CE, individual nucleotides can be identified.
2. **Drug analysis:** Drugs and related compounds can be identified with CE
3. **To study the characteristics of protein:** Proteins are amphoteric in nature, and they can be separated and purified by CE using the isoelectric point pI.
4. **DNA sequencing:** DNA sequencing is the process of determining the order of nucleotides in DNA. This sequencing can be carried out with CE
5. Capillary electrophoresis is also used for the simultaneous estimation of NH_4^+, Na^+, K^+, Mg^{2+} and Ca^{2+} in biological fluids.

CHAPTER 11

Gas Chromatography

Gas chromatography is a most widely used powerful technique in separation chemistry to analyze volatile substances in the gaseous phase. In gas chromatography, gas is used as the mobile phase. If the stationary phase is a solid adsorbent, the technique is termed as gas-solid chromatography (GSC). A liquid on an inert support is used as stationary phase in gas-liquid chromatography (GLC). Thus, based on the stationary phase, the gas chromatography is classified as GSC or GLC. Gas-liquid chromatography is the most commonly used method to separate volatile organic compounds. In identification of molecules, the combination of gas chromatography and mass spectrometry is an invaluable tool.

A GC chromatogram is a representation of the separation of the components that occurs in the GC column. Each of the peaks in the chromatogram represent the signal corresponding to a component eluting from the GC column into the detector. The x-axis shows the Retention time (R_t), and the y-axis shows the intensity of the signal. The chromatogram is plotted by the software in the computer. The area under the peak is proportional to the component concentration and is used for quantitative estimation.

THEORY

When a gas comes into contact with the adsorbent, some quantity of gas gets adsorbed on its surface. Adsorption isotherm is a graph or a relation between the amounts of adsorbate adsorbed on the surface of adsorbent and pressure at a constant temperature. The two important equations in adsorption are Freundlich and Langmuir adsorption isotherms. The **Freundlich equation** or **Freundlich adsorption isotherm** gives the observed experimental relation between quantity of a gas adsorbed onto a solid surface and the gas pressure. The same relationship can be applied to the solute concentration adsorbed onto the surface of a solid and the concentration of the solute in the liquid phase. It is mathematically expressed as

$$x/m = Kp^{1/n} \text{ or } x/m = Kc^{1/n}$$

where

    x = mass of adsorbate

    m = mass of adsorbent

    p = Equilibrium pressure of adsorbate if gas is used as mobile phase.

    C = Equilibrium concentration of adsorbate in solution if the mobile phase is a liquid.

    K and n are constants for a given set of adsorbate and adsorbent.

At high pressure $1/n = 0$, hence extent of adsorption becomes independent of pressure.

Thus, the Freundlich adsorption isotherm fails at higher pressure.

The **Langmuir adsorption model** explains adsorption by assuming an adsorbate behaves as an ideal gas at isothermal conditions. According to the model, adsorption and desorption are reversible processes.

The mass of solute adsorbed per unit weight of adsorbent (m) depends on the concentration (c) of the solute. Langmuir equation is based on the assumption that only a monolayer is adsorbed and there is only one binding site on the adsorbent surface. The mathematical expression is given as

$$m = \frac{K_1 K_2 c}{1 + K_2 c}$$

where

K_1 – No of active adsorption sites /unit wt of adsorbent

K_2 – Affinity of the solute for the adsorbent.

c – Concentration of adsorbate

Langmuir assumed only one binding site but actually there are number of sites on the adsorbent surface and multiple layers of adsorption also take place against his monolayer adsorption assumption.

The law governing the separation in GLC, where the stationary phase is a liquid is Henry's law of partition which states that

x/m = Kc where K is the constant and c is concentration. K values differ with different pairs of vapor adsorbent combinations.

PRINCIPLE

In **GSC**, where the stationary phase is a solid, the principle of separation is **adsorption**. This technique has got limited applications like separation of low molecular weight gases like H_2S, CS_2, CO_2, Rare gases, CO and nitrogen oxides.

Principle of separation in **GLC is partition**. The stationary phase is a liquid which is coated as a thin film over a solid support. The mobile phase used is a gas. The components of the mixture to be separated are converted to vapors. These vapors are carried into the column after mixing with the mobile phase gas. The solubility of the component in the stationary or mobile phase (partition coefficient) determines the rate of elution of the component. The component which is more soluble in the mobile phase travels faster and gets eluted first. Thus the separation is based on the partition coefficient of the component for the given set of stationary and mobile phase combination. Under the given conditions, partition coefficient is unique for each component and no two components can have same partition coefficient.

$$\text{Parition coefficient} = \frac{\text{Concentration of the sample in stationary phase}}{\text{Concentration of sample in mobile phase (gas)}}$$

Criteria for the compounds to be analyzed by GLC

1. **Volatility of the sample:** The mixture to be separated must be volatile in nature so that it can be mixed with the mobile phase.

2. **Thermo stability of the sample:** The sample should be vaporized by heating before entering into the column. If the sample is decomposed while heating, (thermo labile), then the sample cannot be separated and detected. Thus thermostability becomes an important criterion for the compound to be detected by GLC.

INSTRUMENTATION

Components of gas chromatography are

1. Carrier gas
2. Flow regulators /Flow meters
3. Injection devices
4. Columns
5. Temperature control devices
6. Detector
7. Recorders and integrators.

1. **Carrier Gas:** The carrier gas acts as the mobile phase and it carries the sample mixture through the column for separation into its components. The commonly used carrier gases are Helium (He), nitrogen (N_2), hydrogen (H_2), and argon (Ar). The choice of carrier gas is often dependent on the type of detector which is used. The carrier gas system also contains a molecular sieve to remove water and other impurities. **Helium and nitrogen are the most commonly used gases.**

Table 11.1 Components of gas chromatography.

Carrier gas	Flow regulators /Flow meters	Injection devices	Columns	Temperature control devices	Detector
Hydrogen Helium Nitrogen Argon	Rota meter Soap bubble meter	Split injectors Splitless injectors On-column injectors Automatic injectors	Analytical Preparative Packed Capillary WCOT SCOT PLOT	Preheaters Thermostatically controlled oven	**Concentration dependent detectors** Thermal conductivity detector, electron capture detector **Mass flow dependent detectors** FID, FPD, MS, AES and nitrogen phosphorous detectors.

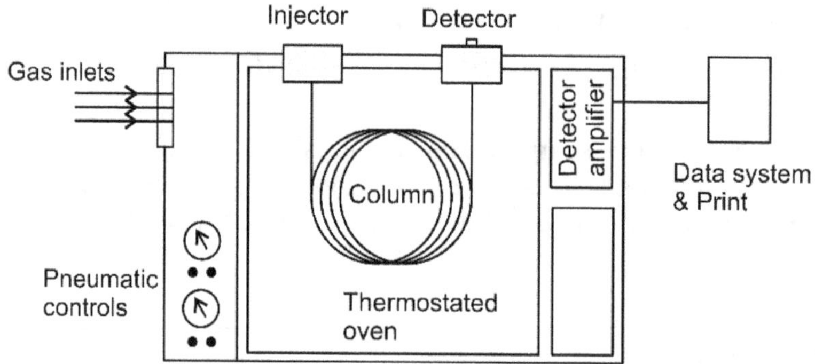

Fig. 11.1 Schematic diagram of gas chromatography.

Criteria for the carrier gas
1. It should be cheap, inert and readily available.
2. It should be of high quality and should not have fire hazard.
3. It should be suitable for the analyte and the detector.
4. It should give good column performance with speed of analysis.

Hydrogen: Hydrogen gas has good thermal conductivity and low density. It can be used with thermal conductivity detector and flame ionization detector.

Disadvantages:
1. It reacts with unsaturated compounds
2. Inflammable

Helium: Helium gas has got excellent thermal conductivity. It is the most commonly used carrier gas. It is inert in nature. It has low density and it allows greater flow rates. It is a safe gas. It has relatively wide optimum linear velocity range.

Disadvantage: Expensive

Nitrogen: It is an inexpensive and safe gas. Its sensitivity is less.

As the carrier gases are compressible, they are stored under high pressure in cylinders.

2. **Flow Regulators and Flow Meters:** The high pressure of the carrier gas cylinders necessitates the use of flow regulators so that gas can be delivered with uniform pressure or flow rate.

Flow meters are used for measuring the flow rate of carrier gas. There are two types of flow meters.

(a) Rota meter
(b) Soap bubble meter

(a) **Rota meter:** It is a calibrated glass tube which has a float held on a spring. The level of the float indicates the flow rate of the carrier gas. It is placed before the inlet of the column.

(b) **Soap bubble meter:** It has a graduated pre calibrated glass tube attached with a inlet tube at the bottom for the carrier gas. The soap solution is stored in a rubber bulb. When this bulb is pressed, a drop of the soap solution gets converted to a bubble as the pressurized carrier gas comes in through the gas inlet. As the gas travels up, it takes the bubble along with it. The distance travelled upwards gives the measure of carrier gas flow rate.

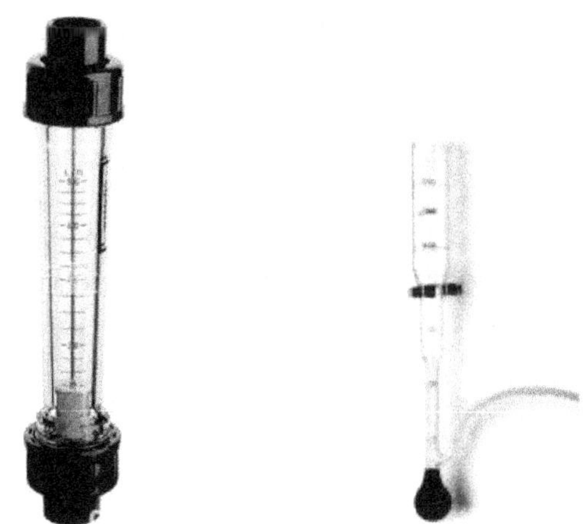

Fig. 11.2 Rota meter.   **Fig. 11.3** Soap bubble meter.

3. **Injection Devices:** The sample is introduced into the column head through the injection device. The injection port is included in a oven that helps to maintain the temperature 20-50°C above the boiling point of the sample, so that the sample can be converted into vapor.

   Gaseous samples are introduced into the column using air tight **hypodermic needle** of 0.5 ml-10 ml capacity through gas sampling valves.

   For **Liquid** samples, **Hypodermic syringes** are used to inject through self sealing rubber septum. (0.1 -100µl capacity)

   Solid samples are **dissolved in volatile solvent** and then **injected** through a septum

   Types of injectors
   (a) Split injectors
   (b) Split less injectors
   (c) On-column injectors
   (d) Automatic injectors

   (a) **Split injectors:** Split injection is the most popular and versatile method in capillary GC analysis. The sample is introduced into the inlet and is vaporized as it moves downstream. Large portions of the vaporized sample have to be exhausted before the sample is separated in the column and detected. The split line has a role to protect the column from any sample overload and increase the moving velocity of the component in the injector port. The sensitivity of the injectors are low but the resolution is not affected.

   (b) **Split less injectors:** The split less injector is suitable for relatively high boiling point compounds with a low analyte concentration. The entire sample reaches the column in these injectors. They are used in high capacity columns such as Support-Coated Open Tubular (SCOT)/heavily coated Wall-Coated Open Tubular (WCOT) columns.

(c) **On-column injectors:** Inlet temperature is kept below the boiling point of the sample solvent and the tip of the micro syringe is inserted directly into the tip of the capillary column for sample injection. Thereafter, by raising the temperature of the inlet and column, the sample is gradually vaporized directly inside the capillary column.

(d) **Automatic injectors:** These injectors are used when large number of samples are analyzed and also for improving reproducibility.

4. **Columns:** Columns are the most important part of the instrument. They are made up of glass, stainless steel, copper, aluminium, nylon and other synthetic plastics. Glass columns are inert but they are fragile in nature.

Classification of columns: columns can be broadly classified into two types

(i) **Analytical columns:** These are packed columns made up of glass or stainless steel. They are about 1-1.5 m length and 3-6 mm in outer diameter. These columns can handle only a small quantity of samples.

(ii) **Preparative column:** They can handle large quantity of sample. They are larger in size with about 3-6 m in length and diameter of 6-9 mm.

Analytical columns are further subdivided into

(a) Packed columns

(b) Capillary columns /Open tubular column/ Golay column

(a) **Packed columns:** These columns are prepared by packing granular stationary phase in metal, aluminium, stainless steel, copper, poly tetra fluoro ethylene or glass tubing. In GSC, the column packing is adsorbents which are size graded or porous polymers. The commonly used adsorbents are Kieselguhr, silica or alumina, hydrated amorphous silica with metal oxides like Fe, Al, Mg, Ca, Na, K and synthetic polymers like squalene, Teflon etc.

In GLC, a liquid is coated over the size graded solid inert support. The inert supports are mostly diatomaceous earths like silica, hydrated amorphous silica with metal oxides or synthetic polymers like squalene and Teflon used in the form of small spherical beads. The stationary phase liquid may be non-polar (e.g. silicone oils-methylsilicone type), polar (polyethylene glycol, silicone oils with cyanopropyl groups) or of intermediate polarity (e.g. phenyl methyl silicone phase, dinonyl phthalate) depending on the type of the separation. The inner diameter of the column ranges from 1.5 mm to 6 mm.

Advantages

1. The porous polymers are stable upto 250°C with no column bleed and no base line drift.
2. Polar compounds are not adsorbed.
3. Higher sample capacity and used for preparative work
4. Porous polymer beads are easily packed in the column and they are mechanically strong.
5. Highly reproducible retention data is obtained.

Micro-packed columns have inner diameter ranging from 0.3-1mm and at lengths from 1-15 m. Particle size ranges from 0.007-0.3 mm in diameter.

The stationary phase is chosen based on the mode of separation and the nature of the analyte.

(a) **Capillary columns:** These are open tubular columns or Golay columns. They are made of long capillary tubing (30-90m) and the internal diameter is 0.10-0.2 mm. The material of construction is stainless steel, copper, nylon or glass etc.

Capillary columns are classified into

(i) WCOT column (Wall coated open tubular column)

(ii) SCOT column (Support coated open tubular column)

(iii) PLOT column (Porous layer open tubular column)

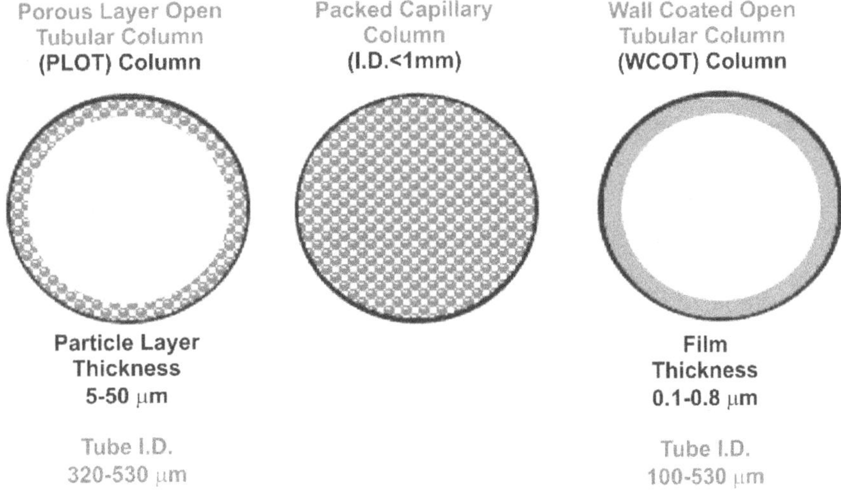

Fig. 11.4 Types of capillary columns.

WCOT column: The inner wall of the capillary column is coated with thin, uniform layer of stationary phase (0.5-1 µ). A very efficient, popular WCOT column is fused silica wall coated open tubular column (FSWC). It is used for the separation of small sample size and it is very inert. There is no resistance offered for the flow of the carrier gas in these columns, because there is no packing in the column.

SCOT column: These columns have a thin layer of (30µm) of supporting material coated on the inner wall of the capillary column and these support material are coated with thin layer of liquid phase. SCOT has more quantity of stationary phase than WCOT and hence can handle more quantity of sample.

PLOT column: This column is similar to SCOT column. The only difference is that the PLOT column docs not have liquid stationary phase. They are used for GSC. The stationary phase coating is 5-50 microns thick.

SCOT columns are less efficient than WCOT columns.

Capillary columns are more efficient than packed column.

If the stationary phase is a solid adsorbent (GSC), the principle of separation is adsorption.

If the stationary phase is a liquid coated on a porous material (GLC),the separation principle is partition.

Equilibration of the column: Equilibration of the column is done before the introduction of sample in packed columns. Equilibration is done by passing the heated carrier gas continuously through the column for a specified period of time (24 hrs) at a particular temperature.

This equilibration helps in getting zero base line when the carrier gas alone is passed through the column.

5. **Temperature controlling devices:** The column temperature can be controlled by jackets containing vapors of boiling liquid, electrically heated metal blocks or circulating air baths. Compounds of low boiling point are eluted at lower temperature. Compounds of high boiling point require temperature programming to avoid broader and shallow peaks.

   The devices used in temperature programming are

   1. Pre heaters
   2. Thermostatically controlled oven

   1. **Pre heaters** are used to vaporize the sample and mix it with the mobile phase or carrier gas. They are present along with the injecting devices.

   2. **Thermostatically controlled oven:** The principle of separation in GLC is partition. The partition coefficient of the sample varies with temperature. Hence the maintenance of temperature in the column and the injection devices is very important for the resolution of the components.   Since the columns are very long they are used in the coiled form and enclosed in the thermostatically controlled oven. The oven can maintain temperature difference up to $0.1°C$.

6. **Detectors:** Detectors are the important part of gas chromatography. The solute particles get eluted along with the carrier gas from the column and enter the detector. The detector produces electrical signals proportional to the concentration of the solute components. These signals are amplified and recorded as chromatographic peaks. In GC, the area under the peak is proportional to the concentration of the analyte.

   Properties of ideal detector

   1. Sensitivity should be high even in small concentrations.
   2. It should be able to detect wide range of samples.
   3. It should give linear and rapid response even at high temperatures
   4. It should be non destructive, simple, easy to maintain and inexpensive
   5. It should be stable and reproducible.

   The Detectors used in Gas chromatography are classified as

   1. Concentration dependent detectors
   2. Mass flow dependent detector

   Concentration dependent detectors: The signal produced is directly proportional to the concentration of solute in the carrier gas. These detectors are **non destructive** detectors.

   Dilution with detector gas lowers the detector response

   E.g., **Thermal conductivity detector and electron capture detector**

Mass flow dependent detector: These detectors are **Destructive in nature** i.e. the sample cannot be recovered. Signal produced is proportional to the rate at which the solute particles enter the detector.

The detector is unaffected by the dilution of carrier gas.

e.g., **FID, FPD, MS, AES** and **nitrogen phosphorous** detectors.

1. **Thermal Conductivity detector/Katharometer:** It is a nondestructive detector. The principle of Katharometer is that **the thermal conductivity of the pure gas changes in presence of the sample component.** (Organic compound)

   Thermal conductivity detector has two platinum wires made of uniform dimensions and they form the two arms of Wheatstone bridge as shown in figure 11.5 b. Through one of the platinum wires, always pure carrier gas flows through and the effluent from the column passes through the other platinum wire. The two platinum wires are electrically heated and they have uniform temperature and electrical resistance.

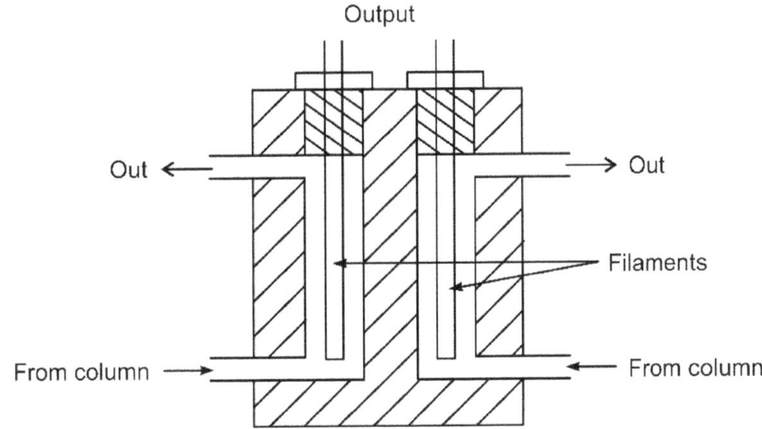

Fig. 11.5(a) Thermal conductivity detector.

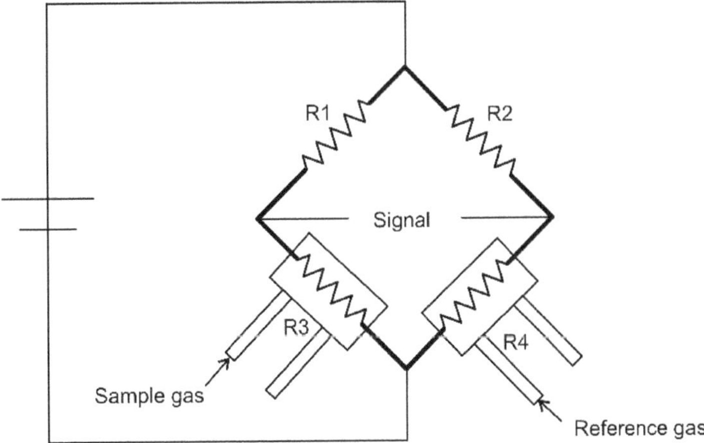

Fig. 11.5(b) Wheatstone bridge.

When only pure carrier gas is passed through both wires, there is no temperature or resistance change which is recorded as base line. When the compound emerges from the column, the wire through which column effluent passes experiences a change in thermal conductivity of the gas leading to change in temperature and resistance of the wire. The difference in the resistance is amplified and recorded as signal.

Hydrogen and helium gas have high thermal conductivity and they are very much suited as carrier gas for TCD.

Advantages

1. Simple and inexpensive
2. Accurate, linear and rugged
3. Applicable to most compounds. Hence called as **universal** detectors
4. **Non –destructive** type. It can be used for **preparative** chromatography.

Disadvantages

1. Sensitivity is low
2. Biological samples cannot be analyzed.
3. Fluctuations in flow rate and temperature affect the outcome.
4. The response obtained is relative response only. Absolute response is not obtained.

2. **Electron capture detector (ECD): The difference in the electron affinity of the substances is used as the principle of ECD.**

   ECD has two electrodes, the anode and the cathode. The column effluent passes between the electrodes. The cathode is treated with a radioactive isotope like ^{65}Ni or ^{3}H emitting electrons as it undergoes decay. When a potential of 20 eV is applied, the secondary electrons are produced by these isotopes. The emitted electrons are collected by the anode producing a constant current.

   If helium is preferred as carrier gas for the column, a make-up gas such as nitrogen or argon/methane is used to increase detector sensitivity. The function of the make-up gas is to improve the detector ionization yield. In other words: make-up gas does not play a role in the chromatographic process in the column. It is only added the moment the carrier gas leaves the column.

   When only carrier gas is flowing between the electrodes, all the secondary electrons are collected by the positive electrode and constant current is recorded as base line. When the carrier gas with the effluent molecules is flowing between the electrodes, the effluent molecules having affinity for the electrons (electronegative components) capture the electrons. This causes a reduction in the constant current. The difference in current is recorded as signal. A schematic diagram of electron capture detector is given in figure 11.6.

   Choice of the carrier gas depends on the electron affinity of the analyte in these detectors. For analytes of high electron affinity, argon gas is used and for low electron affinity analytes nitrogen, helium, hydrogen or carbon di oxide is used.

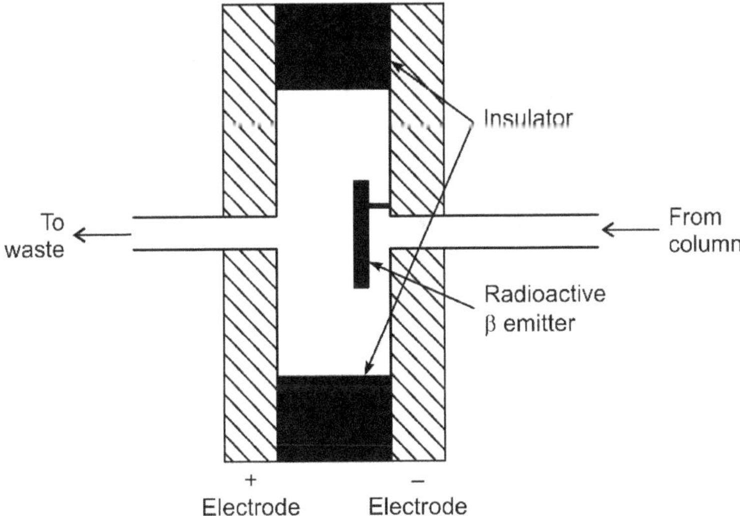

Fig. 11.6 Schematic diagram of electron capture detector.

Advantages:
1. Highly sensitive. Even nanogram quantity of substances can be detected for halogens, nitriles etc.
2. Highly selective
3. Non-Destructive
4. More sensitive than TCD and FID

Disadvantages:
1. Can be used **only for components with electron affinity** eg: Halogenated compounds, Pesticides etc
2. Carrier gas should be pure
3. Insensitive to amines, alcohols and hydrocarbons.

3. **Flame ionization detector (FID):** These detectors are based on the **principle of change in electrical conductivity of carrier gases due to the ions of the components.** At standard temperature and pressure, gases are insulators, but the presence of ions makes them as conductors. In the given conditions if the gas molecules are not ionized, the conductivity of the gas is due to the solute ions present in it.

If a carrier gas other than hydrogen, like nitrogen or argon is used, then they are mixed with hydrogen and they reach the platinum tip of the burner acting as cathode. Silver gauze placed above the burner tip acts as anode. The sample in carrier gas is burnt in air in the detector.

When the carrier gas alone passes through the cathode, there is no ionization taking place and as a result no current is flowing through the system. When a sample component emerges from the column effluent with the carrier gas, the sample gets ionized by the thermal energy from the flame. The collection of these ions by the anode causes a potential difference and current is flowing. This current is amplified and recorded as signal.

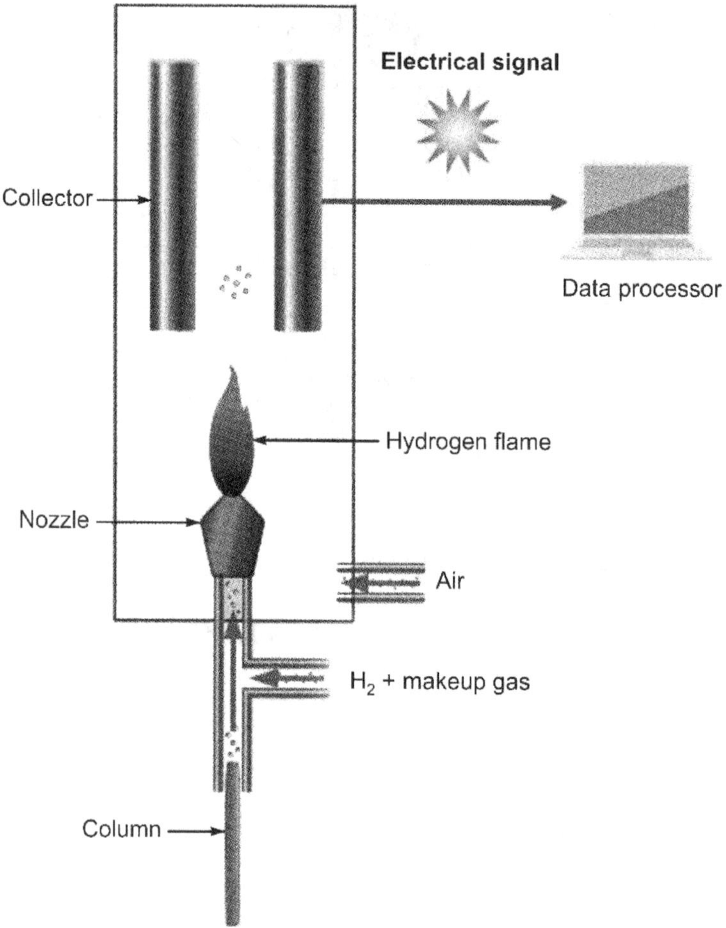

Fig. 11.7 Flame ionization detector (FID).

Advantages

1. Rugged and sensitive. Background noise is low. Hence μg quantities of solute can be detected.
2. Presence of water vapor is not affecting the performance.
3. Responds to most organic analytes except formic acid.
4. Excellent linearity.

Disadvantages

1. **Destructive detector.** Sample recovery is not possible
2. Weakly sensitive to carbonyl, amine and alcohol.
3. Not sensitive to non-combustibles like H_2O, CO_2, SO_2 and these cannot be detected.
4. **Flame photometric detector (FPD):** The thermal energy of the flame in this detector excites the solute particle atoms and the excited atoms emit characteristic radiation which

is detected by the detector. The amount of emitted radiation is directly proportional to the solute concentration in the carrier gas.

This detector is used for the detection of heavy metals like chromium, selenium, tin etc in organo metallic compounds and sulphur and phosphorous containing compounds. Analysis of pesticides, coal, air and water pollutants are also carried out with this detector.

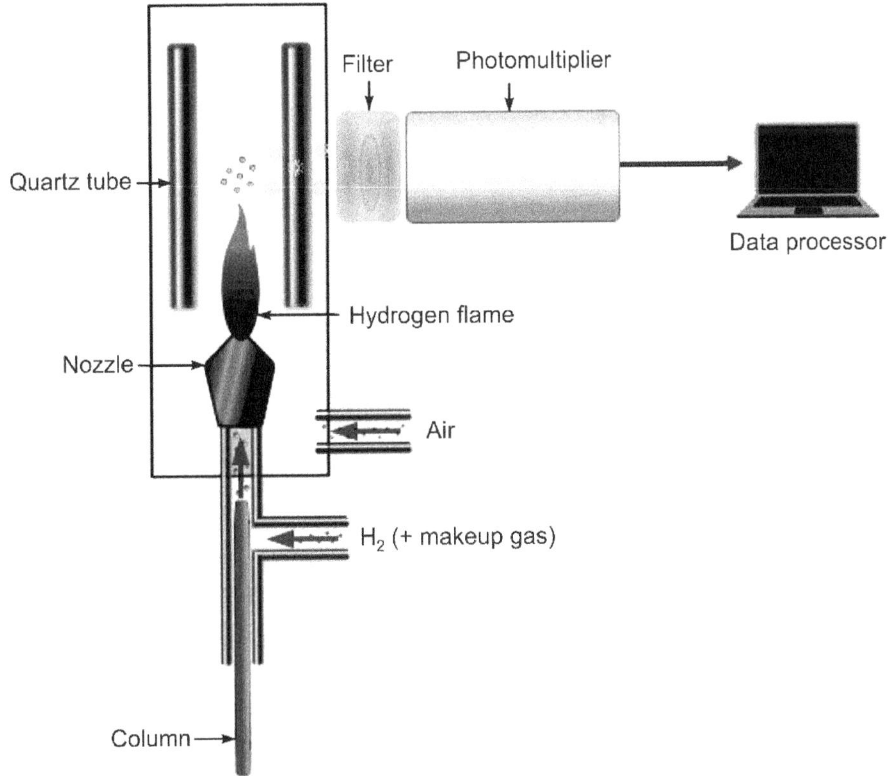

Fig. 11.8 Flame photometric detector.

5. **Argon ionization detector:** Carrier gas used in this detector is argon gas. The argon atoms are excited to a meta stable state using radioactive energy and these argon meta stable molecules collide with solute molecules in the effluent carrier gas and ionize them. These ions are detected by the increase in current

Excitation of argon atoms is carried out with α or β particles. α particles obtained from radium –D, β particles obtained from ^{90}Sr or tritium. These high energy particles are used for the ionization of argon atoms.

Advantages:

1. Most of the organic compounds can be detected.

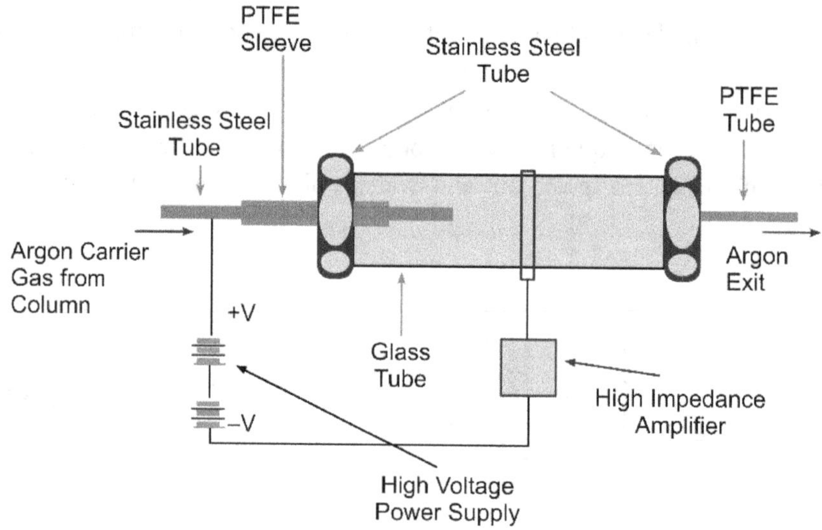

Fig. 11.9 Argon ionization detector.

Disadvantages:

1. Response is relative response not absolute. Response varies with temperature of detector.

2. Poor linearity

3. Less sensitive for halogenated compounds and sensitivity is affected by water vapor.

6. **Thermionic Detector/Nitrogen-Phosphorous Detector (NPD):** A NPD is based on the same basic principles as an FID. A special bead of rubidium is mounted in a silica lattice between the jet and the polarization electrode. This bead is fused to a platinum wire. The bead is heated either electrically or by the flame. Compounds are burned in plasma supplied with hydrogen and air surrounding the rubidium bead. Nitrogen and phosphorous containing compounds produce ions that are attracted to the collector. The number of ions hitting the collector is measured and a signal is recorded. A small amount of alkali metal vapor in the flame greatly enhances the formation of ions from nitrogen and phosphorus-containing compounds.

The NPD is about 500 times more sensitive in detecting phosphorous-containing compounds, and 50 times more sensitive to nitrogen-containing compounds when compared to FID.

Applications: Organophosphate in pesticides and in drug analysis for determination of amino drugs or basic drugs

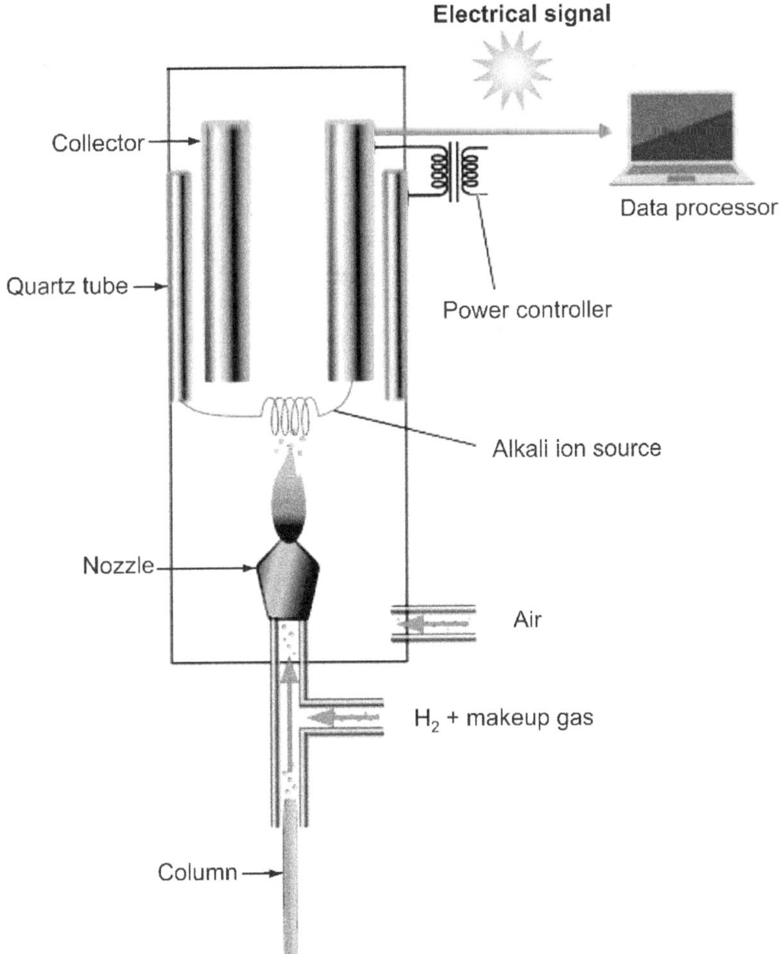

Fig. 11.10 Thermionic Detector/Nitrogen-Phosphorous Detector (NPD).

7. **Electrolytic Conductivity Detector:** It is an element-selective detector for halogen, sulfur and nitrogen containing compounds

   Halogens, sulfur or nitrogen containing compounds are mixed with a reaction gas in a small reactor tube made of Ni. The products from the reaction tube are then dissolved in a liquid, which produces a conductive solution. The conductivity changes due to the ionic species is measured.

   Each detector requires gas, called the detector gas, based on its principle of detection. For example, the flame ionization detector (FID) uses a hydrogen flame so it requires hydrogen and air.

   Analysis using a capillary column requires a makeup gas added just before the detector to act as an auxiliary gas and ensure the detector receives a rapid supply of compounds. Makeup gas reduces the effects of increasing and decreasing column flow rates on detector sensitivity by increasing the sample transfer speed inside the detector and preventing peak broadening. A summary of the detectors in given in the table 11.2.

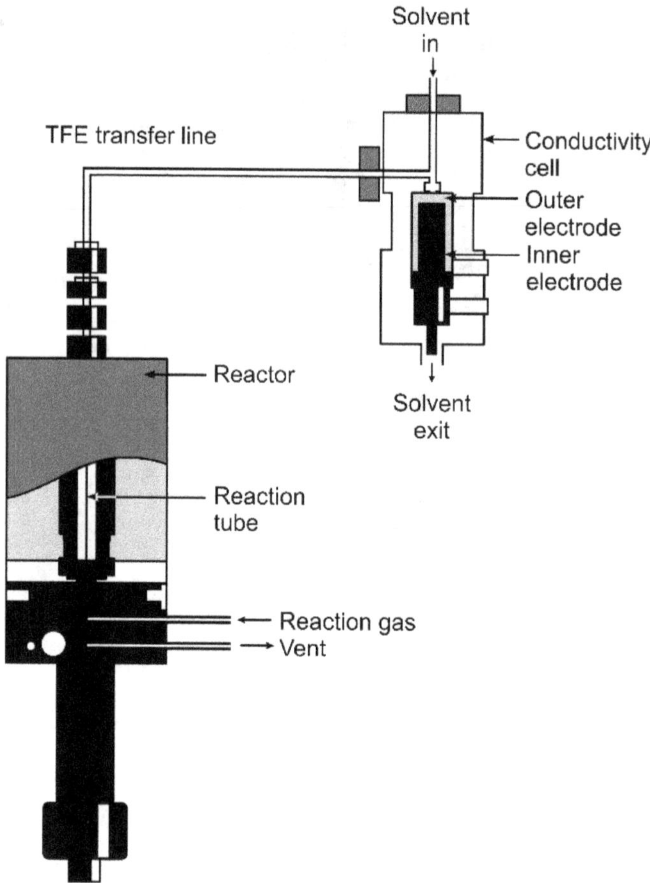

Fig. 11.11 Electrolytic conductivity detector.

7. **Recorders and integrators:** Recorders record the responses from the detector. They record the base line and all the peaks with respect to time.

Integrators: They are the data processors. They record all the data related to every peak like R_t, peak area, peak width and height etc.

Temperature Programming: Temperature programming is controlled increase of temperature during an analysis. In a mixture containing compounds with a wide boiling range, efficient separation within a short analysis time is not possible. In temperature programming, the components with wide boiling range are resolved efficiently in a shorter analysis time and they can be eluted fast with improved peak shapes.

Two types of temperature programming are used

(a) **Isothermal programming:** In this type of temperature programming, the same temperature is maintained throughout the process of isolation. This type of programming is used in separation of components with closely similar boiling points. Adequate separation in short analysis time is obtained at a particular temperature.

Table 11.2 Summary of Gas chromatography detectors.

Detectors	Type	Carrier gas	Make up gas (capillary column)	Selectivity	Sensitivity	Detection limit
Thermal conductivity detector (TCD)	Concentration	He	He or Ar or N_2 or H_2 ,etc.	Universal	Low	10ppm (10 ng)
Flame ionisation	Mass flow	H_2 and Air	He or N_2	Organic compounds (other than formaldehyde and formic acid)	Highly sensitive	0.1 ppm (0.1 ng)
Electron capture detector (ECD)	Concentration	Mainly N_2	Mainly N_2	Halides, nitrates, nitriles, peroxides, anhydrides and organo metallics	Sensitive to electronegative groups	0.1 ppb (0.1 pg)
Flame photo-metric	Mass flow	H_2 and Air	–	Sulphur, phosphorous, tin, boron, arsenic, Germanium, selenium, chromium.	Highly sensitive	10 ppb (10 pg)
Argon ionization detector	Concentration	Argon	–	Organic compounds	Highly sensitive	0.1 ppm

(b) **Linear programming:** There are three types of linear programming

1. Temperature increasing linearly with time
2. Linear increase followed by a period of Isothermal programming
3. Multi linear programming

1. **Temperature increasing linearly with time:** In this method the oven temperature is increased linearly over a period of time. The rate of change of temperature is maintained constant throughout separation. eg: Initial temperature followed by 5°C /minute increase up to the required temperature till the end of separation.

This linear programming is used for the separation of complex mixtures containing low and high boiling point compounds.

2. **Linear increase followed by a period of Isothermal programming:** This linear programming is used for the separation of complex mixtures and the temperature is maintained constant till all the components are resolved.

3. **Multi linear programming:** This programming changes the oven temperature from an initial value to final temperature with various rates, times and temperatures. An example of multilinear programming is shown in figure 11.12

Three variables play important role in temperature programming

(i) **The initial temperature:** The temperature where most of the volatile components are well separated is kept as the initial temperature. It is generally well below the boiling point of the lowest boiling component.

(ii) **The Final temperature:** It is the temperature at which highest boiling component leave the column after separation. The temperature is near the boiling point of highest boiling component in the sample mixture.

(iii) **Rate of rise of temperature:** It can be linear, multi linear, convex or concave.

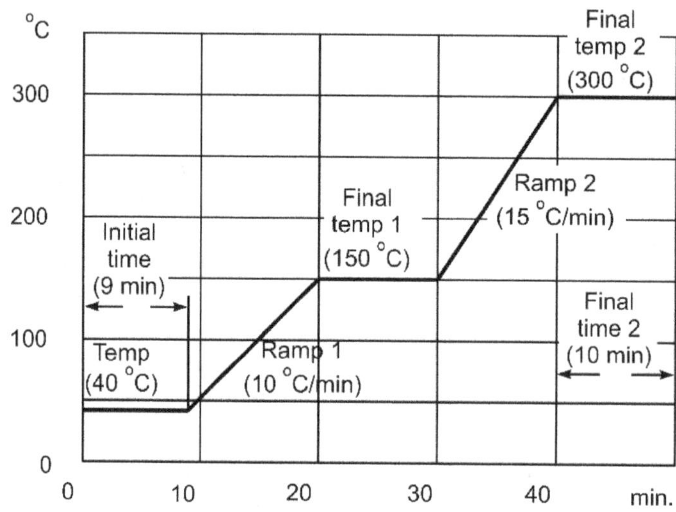

Fig. 11.12 Example of multi linear programming.

Derivatization of sample: It is the technique used to improve the process of separation of the sample by column or detection by detector. **Derivatization is the process of chemically modifying the compound to produce a new compound, so that it can be analyzed by GC.**

There are two types of Derivatization

(a) Pre column derivatization

(b) Post column derivatization

Pre column derivatization: The components are converted to volatile and thermo stable derivatives. Derivatization is performed offline.

Pre column derivatization is done for the following reasons

1. Substance is less volatile or less stable.

2. Components to be separated are thermo labile

3. To reduce tailing and improve the separation factor

Eg: Carboxylic acids, sugars are converted to less polar compounds by treating with reagents like BSA reagent. (Bis tri methyl silyl acetamide reagent)

Post column derivatization: Post column derivatization is performed only for enhancing the detectability of the analytes. This derivatization is done

1. To improve detector response

2. The ionization or electron affinity of the components is increased.

3. This is an on-line detection technique. The flow rate is neither stopped nor altered.

Pretreatment of solid support: Pretreatment of solid support is carried out

1. To overcome tailing.

2. For the separation of non-polar components like esters, ethers etc.

The different methods used in derivatization are

1. Silylation

2. Acylation

3. Per fluoroacylation

4. Alkylation

5. Esterification

6. Condensation

7. Cyclisation

Silylation: Silylation is a chemical reaction in which a reactive hydrogen atom in OH, COOH, SH, NH, CONH, POH, SOH or enolisable carbonyl is replaced with a silyl group, using Trimethyl silyl (TMS).

The silylation helps in the reduction of the polarity of the analyte, increases its stability, and improve the GC behavior of the solute.

Eg. of silylating agents: Hexa methyl disilane(HMDS), Trimethyl chlorosilane (TMCS), Trimethyl silyl imidazole (TMSI), Bis Trimethyl silyl acetamide (BSA).

The functional groups that can be silylated are alcohol,phenol,carboxyl, amine and amide.

Acylation: Acylation is replacing the active hydrogens from an analyte with functional groups like OH, SH, NH CONH, etc with acyl group(R-C=O).

The acylation is also used for reducing polarity and improving the behavior of the analytes in the chromatographic column. Acylation is used to enhance the detectability in GC with ECD and FID detectors.

Common acylating agents are formic acid, acetyl chloride, acetic anhydride, N-Methyl-bis trifluoro acetamide (MBTFA), bis (trifluoro acetamide), trifluoroacetic acid (TFA), propionic anhydride and butyric anhydride.

Alcohols, amines and phenol are derivatized by acylation.

Alkylation: Replacement of the active hydrogens in an analyte with an alkyl (R) group is alkylation and this results in the reduction of polarity of the sample. The functional groups which can undergo alkylation are OH, COOH, SH, NH, or CONH. Alkylation enhances the detectability of the analytes by ECD.

Eg of alkylating agents are Dimethyl formamide, Tetra butyl ammonium hydroxide, Diazomethane.

Carboxylic acids and Phenols are derivatized by alkylation.

Esterification: The esterification reaction is the alkylation of carboxylic acid or the acylation of the alcohol.

Condensation: Ketones and aldehydes present in the sample are derivatized to prevent hydrogen bonding. The agent used is methoxylamine.

Cyclization: Compounds containing two functional groups in close proximity undergo cyclization to form heterocyclic rings.

PARAMETERS USED IN GC

Retention Time (R_t)

It is the difference in time from the point of injection to the peak maxima. Retention time is the time required for the elution of 50% of the component from the column. The unit for R_t is minutes or seconds. The solutes which are not retained by the column travel straight to the detector and t_m is the time taken for the unretained solute to travel through the column. It is shown in figure 11.13

Retention volume (V_r): Retention volume is the volume of carrier gas required to elute 50% of the component from the column. It is a product of retention time and flow rate.

Retention volume = retention time × flow rate

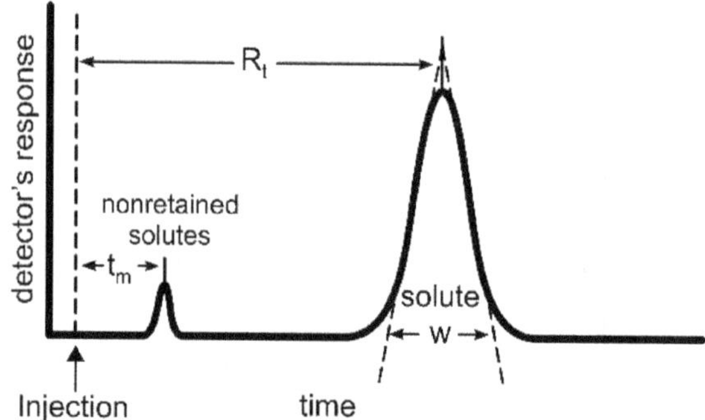

Fig. 11.13 Retention time.

Separation factors(S): It is the ratio of partition coefficient of the two components to be separated. It is expressed as

$$S = K_a / K_b = \left(\frac{t_b - t_0}{(t_a - t_0)} \right)$$

Where

    t_0 = Retention time of unretained substance

    K_b, K_a = Partition coefficients of b and a

    t_b, t_a = retention time of substance b and a

    S = depends on liquid phase and column temperature

If the difference in partition coefficients of two compounds are more, the separation factor is more and the peaks are far apart. If the difference in partition coefficients is less, then the peaks are closer and the separation factor is less.

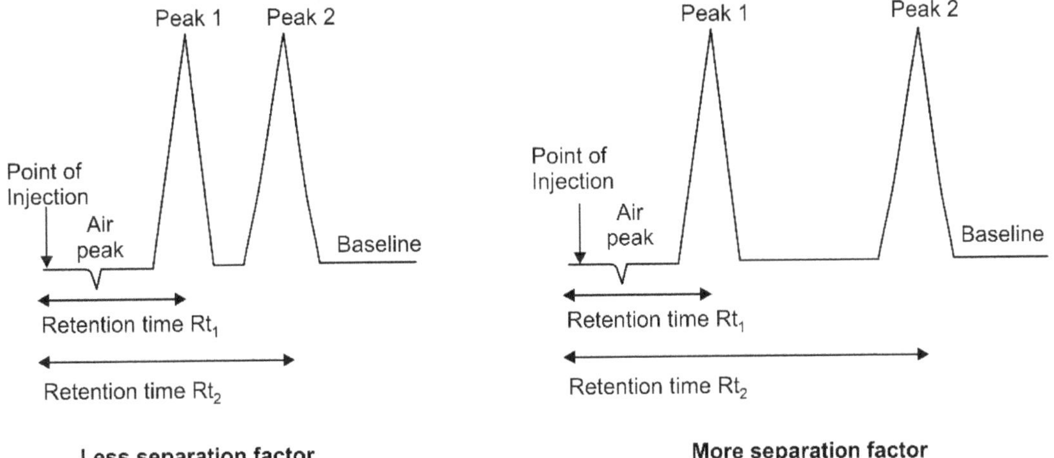

Fig. 11.14 Separation factor.

Resolution: It is a quantitative measure of separation of two components. It is defined as the difference in retention times between the two peaks, divided by the combined widths of the elution peaks. Resolution should be greater than one.

$$R_S = \frac{2(R_{t2} - R_{t1})}{w_1 + w_2}$$

Theoretical plates (Plate theory): This theory was developed by Martin and Synge in 1941. In plate theory the column is viewed as divided into a number of adjacent imaginary segments called as theoretical plates. In each theoretical plate, a complete equilibration of the solute occurs between stationary and mobile phase.

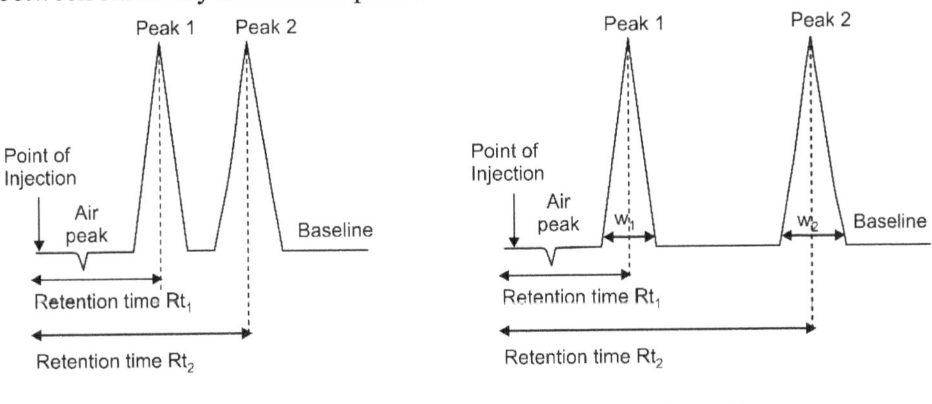

Fig. 11.15 Resolution.

The **(HETP)** Height Equivalent to Theoretical Plates is defined as a unit of column length sufficient to establish equilibrium between the solute in the mobile phase and that in the stationary phase throughout the unit.

The equal interaction of the solute components of the mixture with the mobile phase and stationary phase leads to the retention of the molecules (slow rate of movement).

If each solute molecule is interacting with the mobile phase and stationary phase in a different manner, then they will have different rates of migration leading to separation of the components.

Fig. 11.16 Column showing imaginary theoretical plates.

SIGNIFICANCE OF THEORETICAL PLATES

Efficiency (good separation of the components) of the column increases with

1. Increase in number of theoretical plates. (N)
2. Decrease in HETP (Height equivalent to theoretical plates)

$$H = L/N$$

where

L- Length of the column

N- Number of plates

H – HETP or Height of the theoretical plates.

Van Deemter equation gives the HETP

$$\textbf{HETP} = \textbf{A} + \textbf{(B/u)} + \textbf{Cu}$$

where

A = is the Eddy diffusion

B = Longitudinal diffusion

C = Mass transfer

U - Average mobile phase velocity

Eddy diffusion - (A term) - is caused by the different paths taken by the solute particles in the column resulting in peak broadening. It is independent of flow rate.

Longitudinal diffusion – (B term) – is caused by less analyte concentration in the edges of the column than in the centre leading to peak broadening.

Increased flow rate can control B term and peak broadening can be avoided.

Mass transfer – (C term) – is the movement of the analyte or mass transfer between mobile and stationary phases.

Increased flow rate increases the resistance to mass transfer leading to band broadening.

The Van Deemter plot is shown below in figure 11.17, indicating the desired plate height (H or HETP) and flow rate as optimum efficiency and optimum flow rate respectively.

A column is efficient when the HETP is minimum. Therefore, the flow rate corresponding to minimum value of HETP is used.

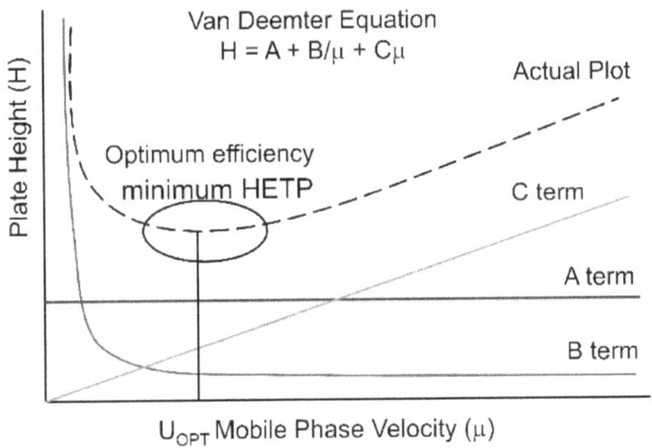

Fig. 11.17 Composite curve of Van Deemter.

Efficiency of the column (No of theoretical plates): The number of theoretical plates in a column represent the efficiency of the column. It is determined using the formula

$$N = 16 \frac{\left(R_t^{\,2}\right)}{\left(W^2\right)}$$

Where   N – no of theoretical plates

       R_t - retention time

       W – peak width at base.

More number of theoretical plates indicate the high efficiency of the column. Gas chromatographic columns normally have 600 - 1000 plates/meter where as in HPLC it varies between 8000- 12000 plates/meter.

Asymmetry factor: A chromatographic peak follows gaussian distribution and it should be symmetrical about its centre. But in actual practice, The peak may not be symmetrical due to some factors, and it shows tailing or fronting.

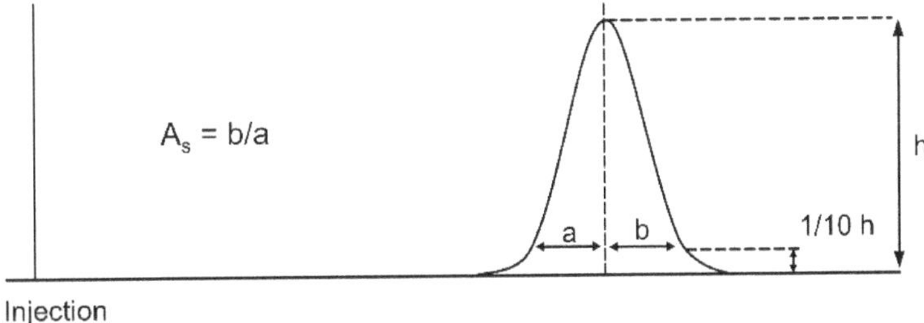

Fig. 11.18 Asymmetry factor.

b - distance from the peak midpoint (perpendicular from the peak highest point) to the trailing edge of the peak measured at 10% of peak height.

a - is the distance from the leading edge of the peak to the peak midpoint (perpendicular from the peak highest point) measured at 10% of peak height.

Asymmetry factor(A_S) should be between 0.95 to 1.05.

If $A_S > 1$ the peak is showing tailing and if $A_S < 1$ fronting is observed in the peak.

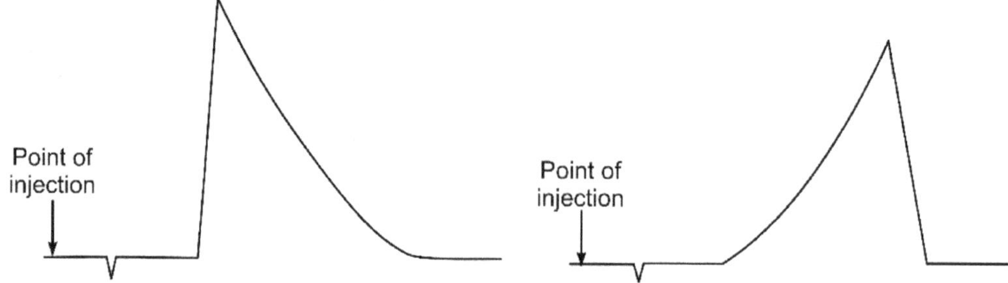

Fig. 11.19 Tailing in a peak.   **Fig. 11.20** Fronting in a peak.

Tailing Factor: Tailing Factor (T_f) is the USP coefficient of the peak symmetry. The formula for tailing factor is given as

$$T_f = (a + b)/2a$$

where a is the distance from the leading edge of the peak to the peak midpoint (perpendicular from the peak highest point) measured at 5% of peak height and b is the distance from the peak midpoint (perpendicular from the peak highest point) to the trailing edge of the peak measured at 5% of peak height.

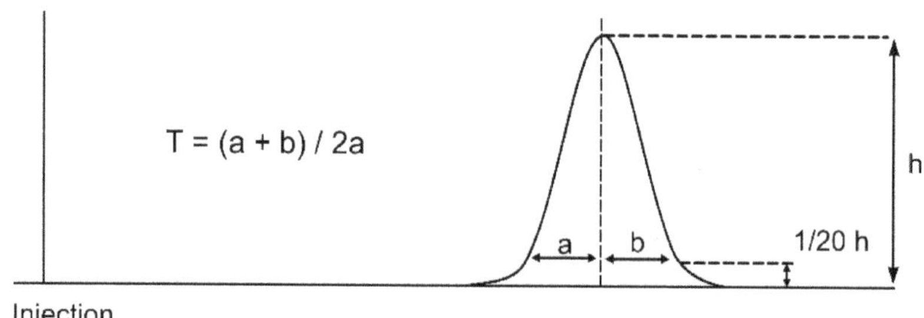

Fig. 11.21 Tailing factor in a peak.

APPLICATIONS

1. **Qualitative analysis:** It is carried out by comparing the retention time of sample with the retention time of the standard. If the operating conditions are same, the retention time also remains the same.
2. **To check the purity of the compound:** Presence of additional peaks in the chromatogram shows the presence of impurities when compared with the standard

chromatogram. The peak areas can be used to calculate the percentage purity of the sample.

3. **Detection of impurities:** Presence of additional peaks when compared with a reference material identifies the impurities. The peak areas can be used to calculate the percentage of impurities.

4. **Quantitative analysis**

   (a) **Direct comparison method:** From the peak areas of the standard and the sample, the quantity of sample can be calculated.

   $$\text{Peak area} = \text{Peak height} \times \text{width at } \tfrac{1}{2} \text{ height}$$

   $$A_1/A_2 = \alpha(w_1/w_2) \text{ where,}$$

   A_1 & A_2 are the peak area of sample and standard.

   α- Response factor

   w_1/w_2 is the weight or concentration of sample and standard.

   (b) **Calibration curve method:** Serial dilutions of standard solutions are prepared and injected in GC. A calibration graph is constructed by plotting peak area vs. concentration. The concentration of the unknown sample is determined from the calibration graph.

   (c) **Internal standard method:** Known concentration of the internal standard (compound with similar retention characteristics) is added to the standard solution and sample solution. The unknown solution concentration is determined using the ratios of sample with internal standard and standard with internal standard. This method is used to avoid the extraction steps when the sample matrix is complex in nature.

5. **Multi component analysis:** GC is also used for the determination of multiple components in a formulation

6. Isolation and identification of drugs in urine, plasma and serum

7. Analysis of mixture of components like amino acid, plant extracts, volatile oils etc.

8. Pharmaceutical applications:

   (a) Quality control and analysis of antibiotics (penicillin), antiviral (amantidine), General anesthetics (chloroform, ether), sedative /hypnotics (barbiturates) etc.

   (b) Assay of drugs like Atropine sulphate etc.

   (c) Presence of related compounds

CHAPTER 12

High Performance Liquid Chromatography (HPLC)

HPLC stands for High-performance liquid chromatography or High-pressure liquid chromatography. It is a type of column chromatography where the mobile phase is pumped with very high pressure into the column through the stationary phase. HPLC yields high performance and high speed compared to traditional column chromatography. It is a chromatographic technique that can be used for the separation and purification of the mixture of components and also for the quantification of the components.

The efficiency of separation in HPLC is very high because of the reduced particle size of the stationary phase (1 to 5μ). This reduction in particle size increases the surface area, which in turn increases the number of theoretical plates. In contrary to this, the decrease in particle size increases the resistance to the flow of the mobile phase. This generates more back pressure in the column, causing damage to the stationary phase, leading to decreased resolution. To overcome these difficulties, high pressure is used to force the mobile phase through the column and hence the name HPLC.

An HPLC chromatogram represents the separation of the components that occurs in the HPLC system. Peaks are rising from a baseline drawn on a time axis. Each peak represents the detector response for a different component. The chromatogram is a two-dimensional plot with the Y-axis (ordinate axis) giving concentration in terms of the detector response, and the X-axis (abscissa) represents the time. The chromatogram is plotted by the computer software. The area under the peak is proportional to the component concentration and is used for quantitative estimation.

TYPES OF HPLC

I. **Based on the principle of separation**
   (a) Adsorption chromatography
   (b) Partition chromatography
   (c) Ion exchange chromatography
   (d) Ion pair chromatography
   (e) Size exclusion/Gel permeation chromatography
   (f) Affinity chromatography
   (g) Chiral chromatography

(a) **Adsorption chromatography:** The separation of components in adsorption chromatography is based on the relative affinities of the components towards the stationary phase. The principle of separation is called adsorption, which is a surface phenomenon. The mixture (adsorbate), which is dissolved in the mobile phase, moves through the stationary phase (adsorbent) of the column. The component with more affinity for the stationary phase travels slower, and the component with less affinity for the stationary phase travels faster down the column. Thus, the components are separated based on their affinity difference for the stationary phase. No two components will have the same affinity for the given set of stationary, mobile phase and other conditions.

If the stationary phase is solid, the principle of separation is adsorption

E.g., GSC, TLC, Column adsorption chromatography, HPLC

(b) **Partition chromatography:** In this mode of chromatography, the separation of components is based on the partition coefficients of the individual components. The solute gets distributed between the two immiscible liquids based on its partition coefficient. The component which is more soluble in the stationary phase travels slower than the component soluble in the mobile phase. The components are separated based on their differences in partition coefficient since no two components will have the same partition coefficient for the given combination of stationary, mobile phase and other conditions.

The stationary phase is a liquid which is coated as a thin film over solid support.

If the stationary phase is a liquid, then the principle of separation is partition.

E.g., Gas-liquid chromatography, paper partition chromatography, column partition chromatography etc.

(c) **Ion exchange chromatography:** The principle involved is a reversible exchange of ions. The stationary phase used is ion exchange resin. For the separation of anions, anion exchange resin is used. For cation separation, cation exchange resin is used. Acids, alkalis and buffers are used as mobile phase.

(d) **Ion-Pair Chromatography:** In this technique, ion-pairing reagents like octane sulphonic acid, Tetraethyl ammonium hydroxide are used to convert an RP HPLC column into a temporary ion exchange column. Ion Pairing agents are ionic compounds that contain a hydrocarbon chain that imparts hydrophobicity to increase the retention of the ion pair. By adding pair ions with the opposite charge to the mobile phase, the pair ions form ion pairs with the ionic compounds, which neutralizes the charge, increasing hydrophobicity and thereby increasing retention.

If pair ions with a hydrophobic functional group (heptafluorobutyric acid in the example shown below) are added to the mobile phase, the hydrophobic functional groups are retained by the stationary phase. The mechanism is shown in figure 12.1.

(e) **Size exclusion/Gel permeation chromatography:** The separation in this chromatography is based on the size of the solute particles. The mobile phase with the sample passes through the gel stationary phase. The gel acts as a molecular sieve and separates the solutes based on molecular size.

E.g., Soft gels: Dextran, Agarose and polysaccharide

Semi-rigid gels: Polystyrene, alkyl dextran.

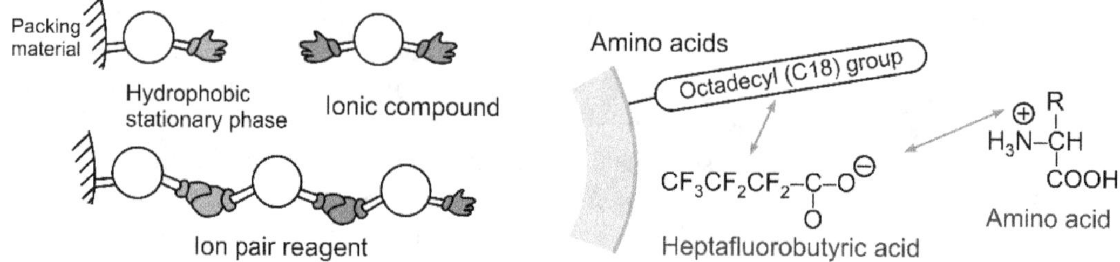

Fig. 12.1 Mechanism of Ion-Pair Chromatography.

(f) **Affinity chromatography:** This chromatography utilizes the specific interaction between the sample and a second molecule immobilized on a stationary phase. Affinity chromatography is mostly used in the field of Biotechnology, Microbiology, biochemistry etc.

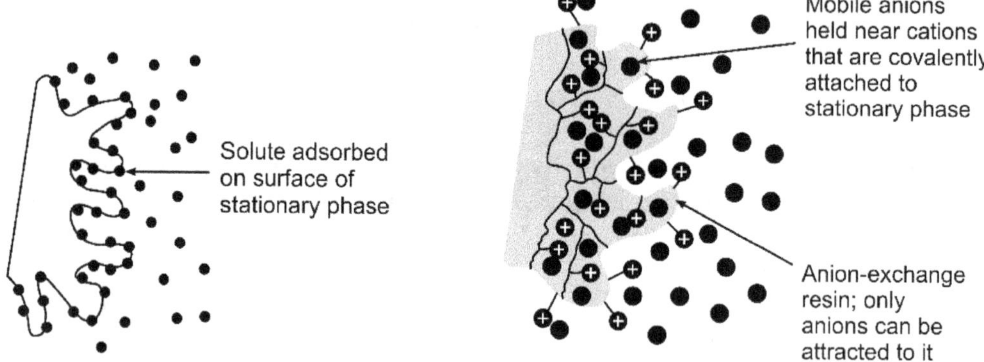

Fig. 12.2 Adsorption Chromatography.   **Fig 12.3** Ion-exchange Chromatography.

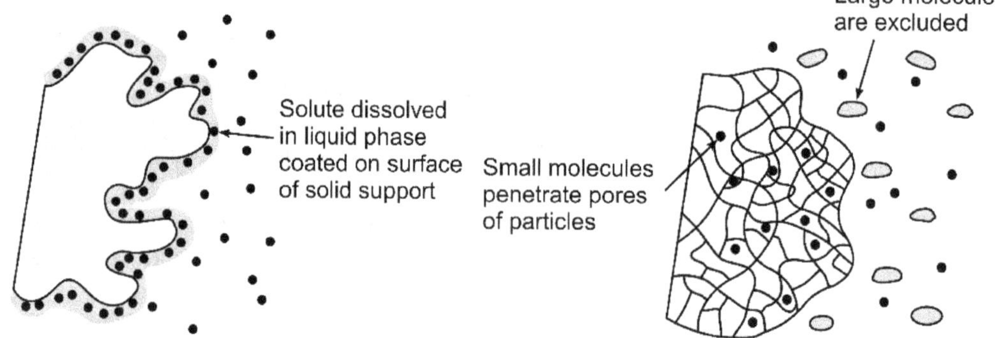

Fig 12.4 Partition Chromatography.   **Fig. 12.5** Size exclusion Chromatography.

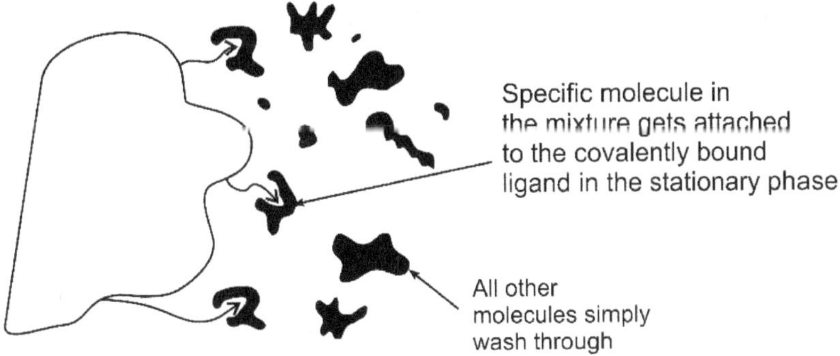

Fig 12.6 Affinity Chromatography.

(g) **Chiral chromatography:** Enantiomers have the same physical and chemical properties and can be separated only in a chiral environment. Separation of enantiomers (optical isomers) is done using either chiral stationary or mobile phase.

II. Based on elution technique

1. **Isocratic elution:** In this separation, the same mobile phase composition and hence the same polarity is maintained throughout the process.
2. **Gradient separation:** A technique in which the mobile phase composition or polarity is changed during the separation process. Linear gradients are the popular techniques for the efficient separation of components with a wide boiling range.

III. Based on the scale of operation

1. **Analytical HPLC:** The **sample size** is **low** in this technique. Qualitative and Quantitative analysis of the sample is carried out using this technique. Recovery of the sample is not possible.
2. **Preparative HPLC:** The **sample size** is **large**. Individual components can be recovered and reused.

IV. Based on the type of analysis

1. **Qualitative analysis:** Qualitative analysis is carried out for the identification of the compounds. It is also useful in finding out the presence of impurities and in finding the number of components.
2. **Quantitative analysis:** This analysis is useful in determining the quantity of the components in the sample mixture by comparing the peak areas of standard and sample.

V. Based on the mode of separation

1. Normal phase chromatography
2. Reversed phase chromatography

PRINCIPLE

The column consists of imaginary plates called as theoretical plates. A theoretical plate is the height of the column in which one equilibration of the solute particle happens with the stationary and mobile phases. When the number of theoretical plates in a column increases, the efficiency of separation also increases. The no of theoretical plates is related to the surface area of the stationary phase. The smaller the particle size more is the surface area. Thus smaller particles increase the efficiency of separation.

But smaller particles offer more resistance to the flow of the mobile phase, causing a back pressure in the column leading to the damage of the stationary phase. Hence the mobile phase is pumped through the column.

The principle of separation in HPLC by both normal phase and reverse phase mode is adsorption. When introduced into the column, the components of the sample mixture travel based on their relative affinities with the stationary phase. The component with more affinity towards the stationary phase (adsorbent) moves slowly, and the component with less affinity for the stationary phase travels faster with the mobile phase.

No two components can have the same affinity for the stationary phase under the given set of conditions, stationary and mobile phases. This forms the basis of separation in HPLC.

The principle of separation in HPLC varies with the type of HPLC. E.g., In partition HPLC, the principle is the partition, in ion exchange HPLC, ion exchange is the principle etc.

Working

1. The prepared mobile phase liquid is poured into the reservoirs after degassing.
2. The mobile phase is pumped into the column under high pressure to get the baseline.
3. The sample is injected through the injection port using injectors and swept into the high pressure liquid carrier stream between the pump and column.
4. As the mobile phase flows through the column, it carries the sample along with it, and the components are separated based on their affinities towards the stationary phase.
5. The detector detects the components as they emerge from the column.
6. The detector signals are amplified and recorded by the detector. The chromatogram contains detector response as a function of time.
7. The components are identified based on their retention time values compared to the standard, and the quantification is done using the peak area.

INSTRUMENTATION

COMPONENTS OF HPLC INSTRUMENT

1. Solvent delivery system
2. Sample injector system
3. Column
4. Detectors
5. Recorders and integrators

1. Solvent delivery system

The components are

(a) Solvent reservoir
(b) Pumps

(c) Mixing units

(d) Gradient controller

(e) Degasser

(f) Dampner

Table 12.1 Summary of HPLC components.

Solvent delivery system	Sample injector system	Columns	Detectors
Solvent reservoir Pumps 1. Reciprocating piston pump, 2. Syringe pump 3. Constant pressure pump Mixing units Gradient controller Degasser Dampner	(i) Septum injector (ii) Stop flow injector (iii) Rheodyne injector	1. Guard column 2. Analytical column 3. Capillary column 4. Fast column 5. Preparatory column 6. Chiral column	1. UV/Visible Detector – Fixed Wavelength – Variable Wavelength – Photo Diode Array (PDA) 2. Refractive index – Detector 3. Fluorescence Detector 4. Electrochemical Detector 5. Conductivity Detector 6. Evaporative light scattering (ELSD) detector 7. Mass detector (LC-MS) 8. Optical rotation detector (chiral detectors)

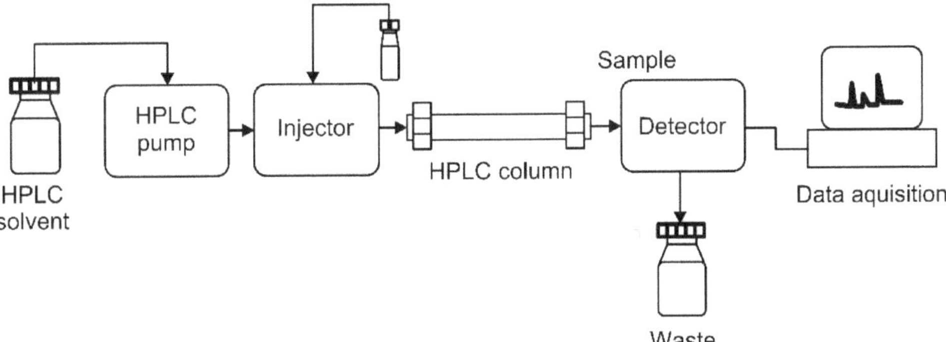

Fig. 12.7 Schematic diagram of HPLC instrument.

(a) **Solvent reservoir:** They are used for storing the mobile phase. They are usually made of glass and they should be kept above the level of the pump. Depending on the mobile phase composition (single solvent or mixture of two solvents), single or multiple solvent reservoirs are used.

(b) **Pumps:** The purpose of the pump is to maintain a constant flow of mobile phase through the HPLC system. This is accomplished regardless of the backpressure caused by the flow resistance of the HPLC column. The pump is used to force the mobile phase through the column at a specific flow rate, expressed in ml/min. Normal flow rates in HPLC are maintained in the range of 1 to 2 ml/min. Pumps can reach pressures up to the range of 6000-9000 psi (400- to 600-bar).The pumps can be used for both isocratic or gradient elution.

Types of HPLC pumps: The most commonly used pumps in HPLC are

1. Reciprocating piston pump
2. Syringe pump
3. Constant pressure pump

1. Reciprocating piston pumps: This pump consists of a piston driven by a small motor. The piston moves rapidly back and forth in a hydraulic chamber of volume 35-400ml.

The piston pulls solvent from the mobile phase reservoir on the backstroke and the separation column valve is closed.

The pump pushes solvent from the reservoir into the column on the forward stroke and the inlet valve is closed.

Flow rates can be changed by altering the piston stroke volume or the stroke frequency during each cycle.

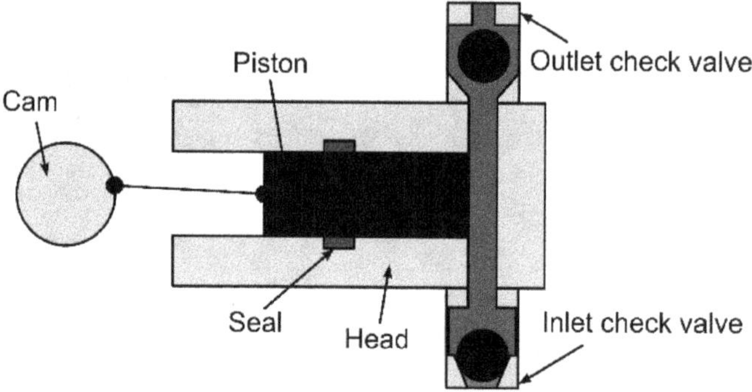

Fig. 12.8 Reciprocating Piston pumps.

An identical piston chamber unit with dual and triple head pump operates at 180 or 120 degrees out of phase (this system is significantly smoother because one pump is filling while the other is in the delivery cycle).

Disadvantages:

1. It produces a pulsed flow which is damped because pulses appear as baseline noise on the chromatograph.
2. Dual pump heads or elliptical cams are used to minimize such pulsations.

Advantages:

1. These pumps have a small internal volume of 35-400ml
2. It can give higher output pressures up to 10,000 psi.
3. Easily adapted to gradient elution.
4. It has got large solvent capacities & constant flow rates.
5. It is largely independent of column backpressure & solvent viscosity.

2. **Syringe type pump/Displacement pumps:** This pump delivers only a finite volume of the mobile phase. This pump has a volume between 250 to 500ml. These pumps are most suitable for small bore columns.

   A motorized screw delivers the mobile phase to the column at a constant rate in this pump. The rate of solvent delivery is controlled by changing the voltage on the motor.

   Advantages:

   The flow is independent of viscosity & backpressure.

   Disadvantages:

   It has a limited solvent capacity & poses a problem when solvents must be changed.

3. **Constant pressure pump:** In these types of pumps, the mobile phase is driven through the column with the use of pressure from the gas cylinder. A low-pressure gas source is needed to generate high liquid pressures. The rapid refill of the solvent chamber is achieved by the valve arrangement. These pumps provide continuous phase flow rates. It is not a constant flow pump. But it is a constant pressure pump. It can deliver high pressures. This pump provides a stable flow during the delivery stroke and stops flow on refill stroke. The cost is low.

   This type of pump provides pulse-less & continuous pumping with high flow rates. It is inconvenient for solvent gradient elution.

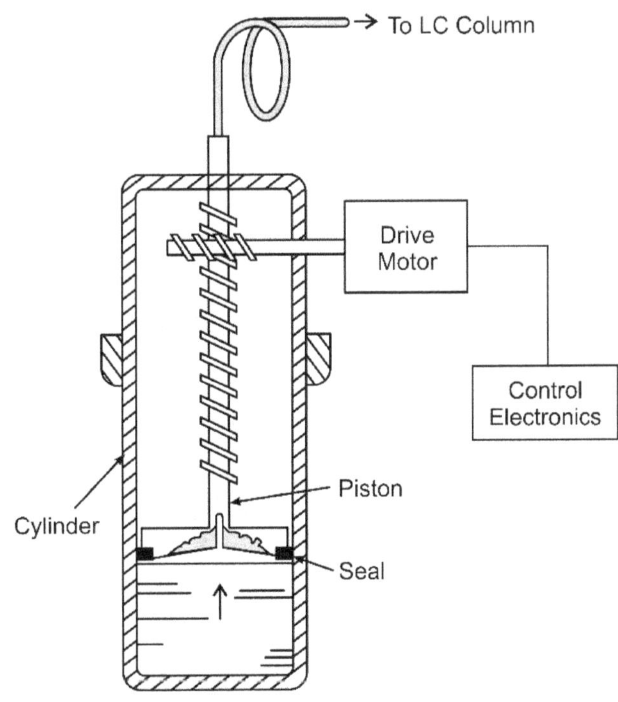

Fig. 12.9 Syringe pumps.

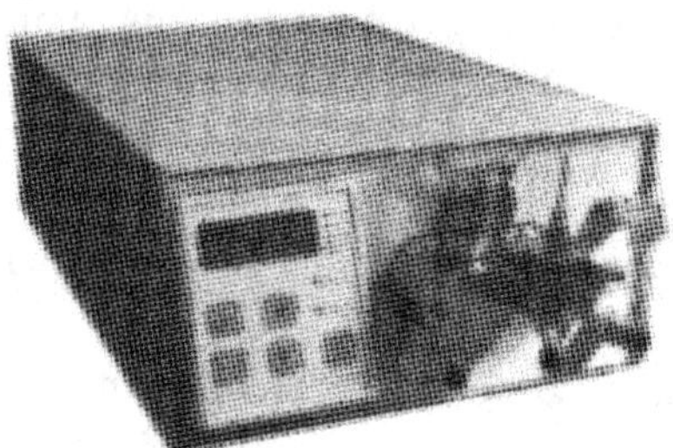

Fig. 12.10 Constant pressure pump.

(c) **Mixing unit:** The mixing unit is used for mixing the solvents in the required proportion and passing it through the column. The low-pressure mixing chamber uses helium for degassing solvents. In the high-pressure mixing chamber, there is no need of helium for degassing solvents.

The mixing unit can be a static mixer or a dynamic mixer.

(i) Static mixer: is packed with beads

(ii) Dynamic mixer: uses magnetic stirrer under pressure.

(d) **Gradient controller:** It is used in the gradient elution technique. The polarity of the solvent is increased gradually, and the solvent composition should be changed accordingly. This is achieved by two or more pumps.

(e) **Solvent degasser:** Organic solvents contain dissolved gases. When the solvents are pumped under high pressure, gas bubbles will be formed, and they interfere with the separation process, baseline steadiness and shape of the peak. Hence degassing of the solvent is necessary, and it is carried out by the following processes

    1. **Vacuum filtration:** This method is accomplished by applying a partial vacuum to the solvent container. This method is not reliable.

    2. **Purging with helium gas:** This method is carried out by passing helium gas through the solvent. It is an effective method. This is an expensive method.

    3. **Ultrasonication:** This technique is carried out using an ultrasonicator. This converts the ultrahigh-frequency to mechanical vibration leading to the removal of air bubbles.

(f) **Dampeners:** Pumps cause a wavy baseline containing pulses. These pulses are dampened by pulse dampeners.

  2. **Sample injector system:** The injector devices available are either for manual or auto-injection. The devices are

(i) Septum injector

(ii) Stop flow injector

(iii) Rheodyne injector

    (i) **Septum injector:** These injectors are used to inject the sample through a rubber septum. They are not commonly used because the septum has to withstand high pressure.

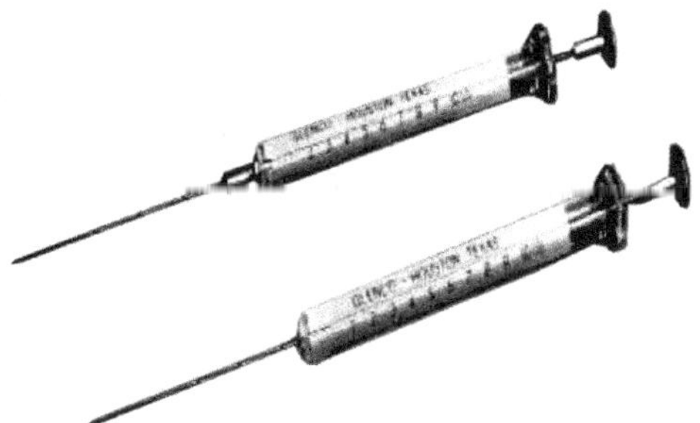

Fig. 12.11 Septum injector.

Disadvantage: The mobile phase leaches the septum, which results in ghost peaks.

(ii) **Stop flow injector (on-line):** In this type of injector, the mobile phase flow is stopped for some time, and the sample is injected through a valve.

Disadvantage: Formation of ghost peaks.

(iii) **Rheodyne injector (loop valve):** It is the most popular injector. The loop has a fixed volume of 20µl or 50µl. There are two modes in the injector.

(i) Load position

(ii) Inject position

The injector introduces the liquid sample into the flow stream of the mobile phase for analysis. It is equipped with six-port valves so that the sample can be injected into the flow path at continuous pressure. For a manual injector, the knob is manually operated to deliver the sample to the column.

The knob is set to LOAD position for sample injection using a syringe, the sample is injected into the sample loop, which is separated from the flow path

The knob is turned to INJECT position, and the eluent travels through the loop from the pump and delivers the sample to the column. Rheodyne injectors positions are shown in Figures 12.12(a) and 12.12(b).

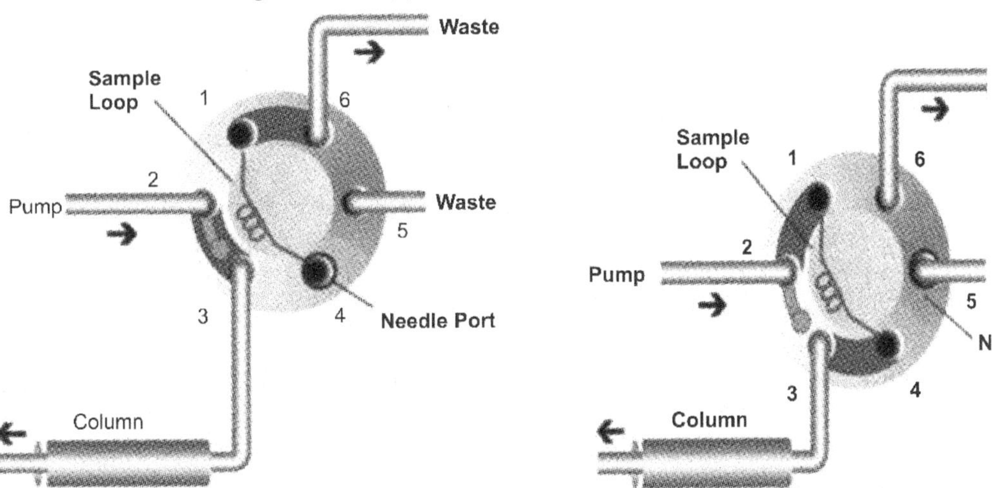

Fig. 12.12(a) Rheodyne injector-load position,   **Fig. 12.12(b)** Rheodyne injector-inject position.

Typical sample volumes for manual injectors are 5-to 20- μL. The injector must also be able to withstand the high pressures of the liquid system. An auto sampler is an automatic version for analysing many samples or when the manual injection is not practical. It can continuously inject a variable volume of 1 μL – 1 mL

3. **Columns:** Columns are the most important part of the instrument, and it is where the separation of components takes place. The columns are packed with the stationary phase through which the mobile phase is flowing through.

The various types of columns used in HPLC are
(i)   Guard column
(ii) Analytical column
(iii) Capillary column
(iv) Fast column
(v)   Preparatory column
(vi) Chiral column

(i) Guard column: This column is introduced before the analytical column, and it improves the life of the analytical column. Guard column packing is the same as that of the analytical column, but the particle size is larger. Guard column acts by
(a) By removing the particulate matter and contaminants from the solvents.
(b) By excluding the components that bind irreversibly to the stationary phase.
(c) It saturates the mobile phase with the stationary phase, so that loss of stationary phase in analytical column is minimized.

(ii) Analytical column: The analytical column is an important part of the HPLC instrument. This column decides the efficiency of separation.LC columns are usually constructed from smooth bore stainless steel tubing or glass. Glass columns are restricted to pressures lower than 600psi.Other materials used for column construction are polyethylene and PEEK(Polyether ether ketone). Column length is about 10 to 30 cm, and the internal diameter is about 4 to 10 mm.

Column packing is usually silica gel. Because of its particle shape, surface properties and pore structure, it gives a good separation. Other materials used are alumina, polystyrene divinyl benzene or an ion exchange resin

Mainly, there are three types of column packing particles; microporous particles, Pellicular particles and bonded phase particles.

Pellicular particle: These are also called superficially porous particles with a particle size of 40 μm in diameter. The pores are coated on an inert core. The inert core is made up of glass or some other hard material. The porous material is made up of polymethacrylates or polystyrene-divinylbenzene. These particles have high mechanical stability and excellent pH stability (pH 2-12).

Porous particle: Porous particles are of 5-10μm in diameter. They are made of silica or aluminium and contain microscopic pores running through the particle to increase the surface area for more interaction with analytes and the mobile phase. E.g., Porous polyvinyl chloride.

Bonded phase particles: These particles also contain an inert core made up of silica. The stationary phase is chemically bonded on that inert core by different linkages like silicate ester, silicate carbon and siloxanes.

E.g., Silica C-18 for reverse phase chromatography of peptides/proteins.

The common columns used in RP-HPLC are C_{18} – Octadecyl silane (ODS) column, C_8- Octyl column, C_4 – Butyl column, CN- Nitrile, NH_2 – Amino column.

(iii) Capillary column: Three types of capillary columns are available. Open tubular, wall coated and SCOT columns. These columns are discussed in GC. The advantages of these columns are decreased sample size(nl), decreased flow rate, decreased solvent volume usage with efficient separation.

(iv) Fast column: These are shorter columns packed with particles of 3µm in diameter. Their advantages are increased sensitivity, decreased analysis time and increased reproducibility.

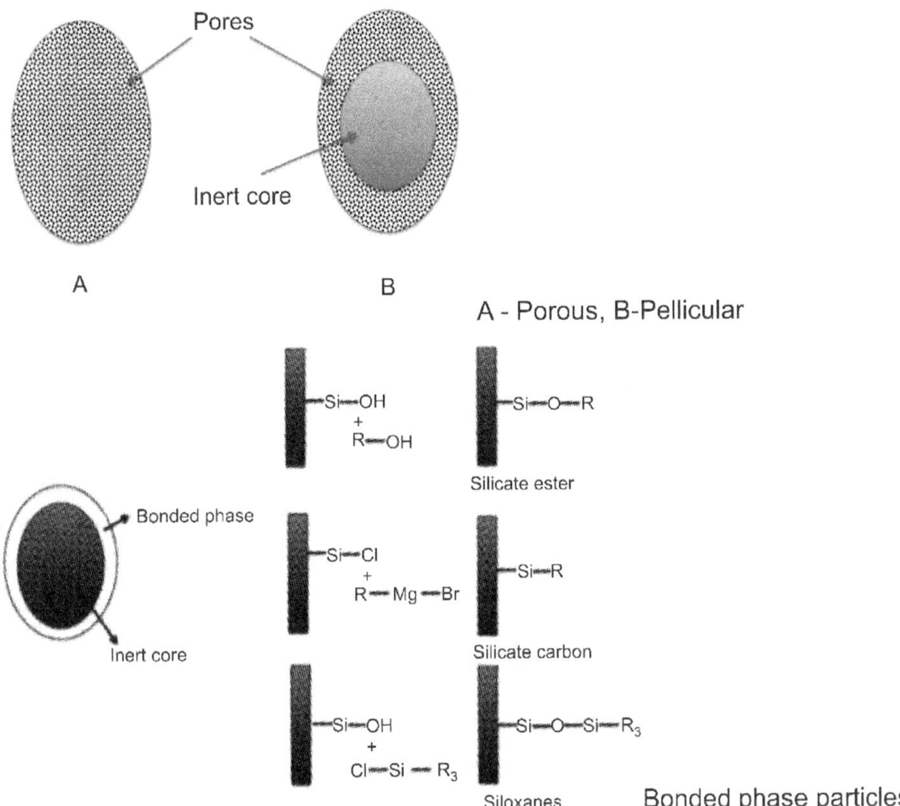

Fig. 12.13 Types of column packing material.

(v) Preparatory column: The length of these types of columns is 10-15 cm and the internal diameter is 25 mm. The particle size used is 15-100µm. They are bigger in dimensions and can handle more sample quantities. They are used in preparatory applications.

(vi) Chiral column: These columns are used in the separation of enantiomers. Chiral columns contain chiral stationary phases.

4. **Detectors:** Detectors are used to detect components eluting from the column by producing an electrical signal. The output of the detector is connected to a recorder or computer.

The selection of the detector is based on the property of the compounds to be separated.

Commonly used detectors in HPLC

There are two types of detectors used in HPLC

1. **Selective detectors (solute property):** These detectors respond to a particular physical or chemical property of the solute and are independent of the mobile phase. E.g., Absorbance detectors, Fluorescence detectors, electrochemical detectors, Mass spectrometric detectors.

2. **Universal detectors (bulk property):** These detectors measure the difference in the physical property of the mobile phase due to the presence of solute particles compared to the mobile phase alone. They are universal in application. They have poor sensitivity and limited range. Such detectors are usually affected by even small changes in the mobile-phase composition. Hence they cannot be used in gradient elution. E.g., Refractive index detectors, Evaporating light scattering detectors.

The various detectors used in HPLC are

1. UV/Visible Detector
2. Refractive index - Detector
3. Fluorescence Detector
4. Electrochemical Detector
5. Conductivity Detector
6. Evaporative light scattering detector (ELSD)
7. Mass detector (LC-MS)
8. Optical rotation detector (chiral detectors)

1. **UV–VISIBLE DETECTORS:** It is a photometric detector. Substances that absorb UV light can be detected by this solute property detector. These are versatile detectors with the best sensitivity and linearity. They are cost-effective, popular and widely used. There are three types of UV detector

   (a) Fixed wavelength detectors
   (b) Variable wavelength detectors
   (c) Photodiode array detectors

   (a) **Fixed Wavelength detector:** Detector operates at 254 nm only because most of the compounds absorb at this wavelength

   Simplest and cheapest of the UV/VIS detectors

   Limited in flexibility

   Limited in types of compounds that can be monitored

   (b) **Variable Wavelength detector:** In this detector, a single wavelength is monitored at any given time but operates in the wavelength range 190-900 nm.

More expensive requires more advanced optics

More versatile, used for a broader range of compounds

More sensitive due to photomultiplier tube

(c) **Photo Diode Array Detector:** In this detector, the absorbance of solutes is measured simultaneously at several different wavelengths. Light from the emission source, such as a deuterium lamp, is collimated by an achromatic lens system. The total light passes through the sample cell and then to a holographic grating. In this way, all wavelengths of light emitted by the source pass through the sample. The dispersed light from the grating is allowed to fall onto a diode array. The array contains many hundreds of diodes, and the output from each diode is processed by a computer and stored. The working of the PDA detector is shown in Figure 12.14.

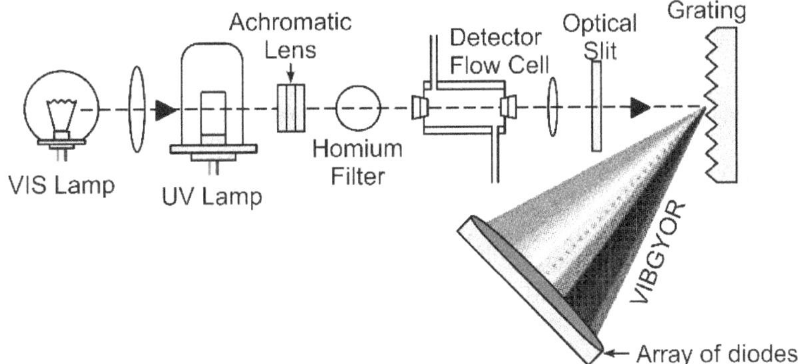

Fig. 12.14 Photo Diode array Detector.

2. **Refractive index detector:** A Refractive index detector measures the molecule's ability to deflect light in a flowing mobile phase in a flow cell relative to a static mobile phase contained in a reference cell. The amount of deflection is proportional to the concentration of the solute in the mobile phase. The refractive index (RI) detector is one of the least sensitive LC detectors. This detector is extremely useful for detecting the compounds that are non-ionic, not showing ultraviolet light absorption and substances that do not exhibit fluorescence. E.g. sugar, alcohol, fatty acid and polymers.

3. **Fluorescence Detector:** These detectors are used for the substances that give fluorescence when subjected to radiation from a source. The excitation light is usually a low-pressure mercury lamp providing relatively high-intensity UV light at 253.7 nm. Many substances that fluoresce will be excited at this wavelength. The excitation light is focused by a quartz lens through the cell. A second lens, set normal to the incident light, focuses the fluorescent light onto a photocell.

As it is very sensitive, its response is only linear over a relatively limited concentration range.

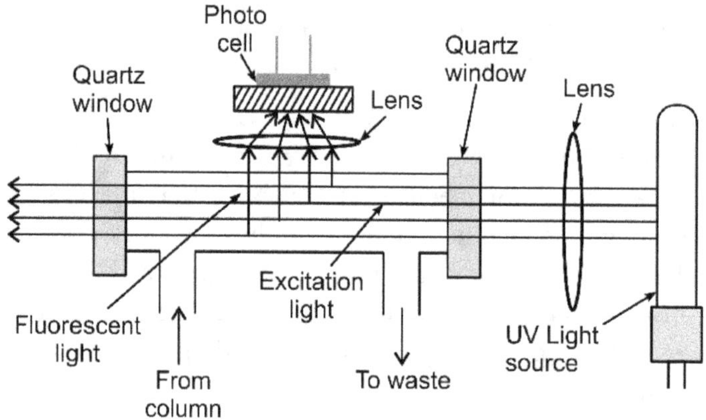

Fig. 12.15 Fluorescence detector.

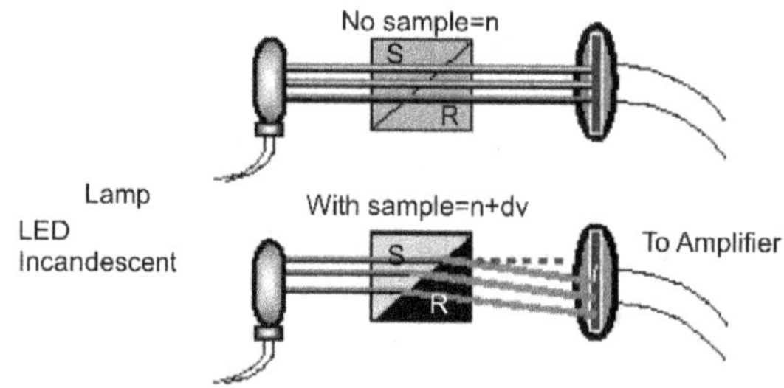

Fig. 12.16 Refractive index detector.

4. **Electrochemical detectors:** The current resulting from an oxidation/ reduction reaction of the analyte at a suitable electrode is measured in this detector. The current produced is directly proportional to the analyte concentration. The eluent should be electrically conductive.

   E.g. of drugs detected using this detector are phenol, catecholamines, nitrosamines, and organic acids in the ng range

5. **Conductivity detectors:** This detector is used in ion-exchange chromatography for the detection of ionic compounds. The ability of the mobile phase to conduct current when placed in a flowcell between two electrodes is measured. Conductivity detectors measure the resistance, which is directly proportional to the concentration of ions present in the solution. Two electrodes placed in the mobile phase form each corresponding to one arm of a Wheatstone Bridge.

6. **Evaporative light scattering detector (ELSD):** Detection is based on the scattering of a beam of light by particles of the compound.
   The three steps involved in detection are
   - Nebulisation
   - Desolvation
   - Detection

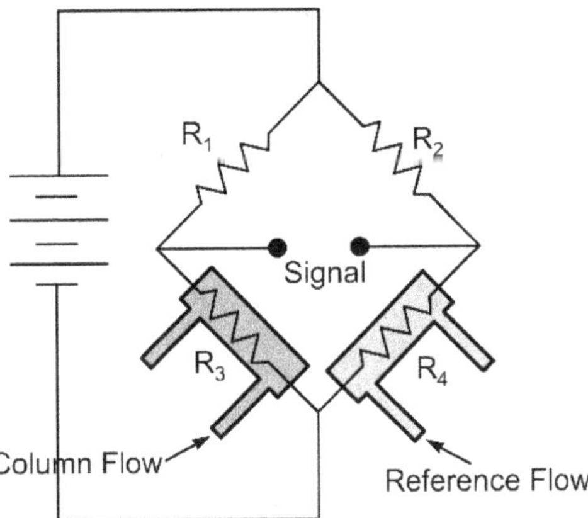

Fig. 12.17 Wheatstone Bridge arrangement in conductivity detectors.

A stream of inert gas nebulises the flow from the column. (Nebulizing is to convert a liquid into a fine spray or mist). The volatile mobile phase is evaporated, leaving tiny particles of the analytes. The particles, when passed through a laser beam, they scatter the laser light. The scattered light is measured at right angles to the laser beam by a photodiode detector. It is a universal detector. The compound does not require a chromophore for detection by this detector.

7. **Mass spectrometer:** The HPLC system in which the Mass spectrophotometer is used as the detector is called LC-MS. In this method, the separation power of HPLC is combined with the detection power of Mass spectrometry. Mass spectrometry (MS) is a powerful analytical tool that can supply both structural information about compounds and quantitative data relating to mass.

8. **Chiral detectors:** Optically active compounds such as amino acids, sugars, terpenes and other compounds containing asymmetric carbon are detected by chiral detectors. There are two types of chiral detection techniques,
    1. Polarimetry or optical rotary dispersion (ORD)
    2. Circular dichroism (CD).
    ORD detectors are based on differences in refractive index
    CD detectors identify the enantiomers by measuring the differential absorption of right and left-handed circularly polarized light.

5. **Recorders and integrators:** Recorders record the responses from the detector. They record the baseline and all the peaks with respect to time. Integrators are the data processors. They record all the data related to every peak like R_t, peak area, peak width and height etc. Computers are used for data processing.

Advantages of HPLC:
1. Speed
2. Efficiency
3. Accuracy
4. Versatile and extremely precise when it comes to identifying and quantifying chemical components.

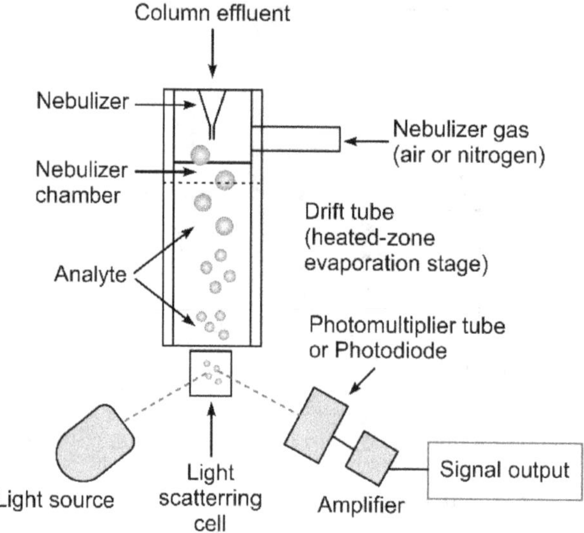

Fig. 12.18 Evaporative light scattering detector (ELSD).

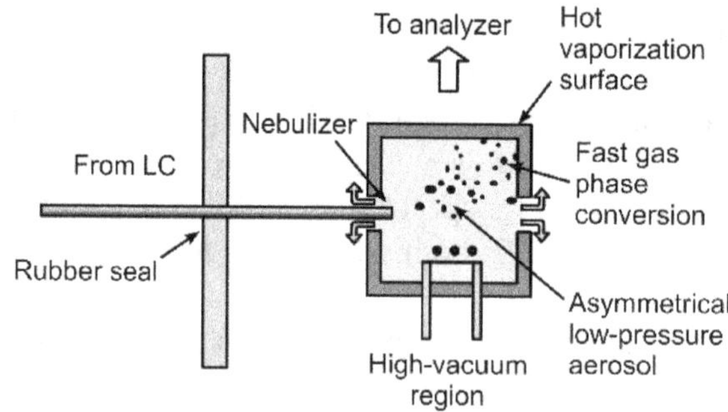

Fig. 12.19 Mass detector.

Disadvantages of HPLC:
1. **Cost:** Costly, requiring large quantities of expensive organic solvents.
2. **Complex to operate:** The compounds that are irreversibly adsorbed onto the stationary phase of the column cannot be detected.
3. Volatile substances are better separated by gas chromatography.

APPLICATIONS

1. **Qualitative analysis (identification):** It is carried out by comparing the retention time of the sample with the retention time of the standard. If the operating conditions are the same, the retention time also remains the same. It is used in fields like environmental, Pharmaceutical, food, forensic, nutraceuticals and cosmetic industries
2. **To check the purity of the compound:** The presence of additional peaks in the chromatogram shows the presence of impurities when compared with the standard

chromatogram. The peak areas can be used to calculate the percentage purity of the sample.

3. **Detection of impurities:** The presence of additional peaks, when compared with reference material, identifies the impurities. The peak areas can be used to calculate the percentage of impurities.

4. **Quantitative analysis**

   (a) **Direct comparison method:** From the peak areas of the standard and the sample, the quantity of the sample can be calculated.

   Peak area = Peak height X width at ½ height

   $A_1/A_2 = \alpha(w_1/w_2)$ where,

   A_1 & A_2 are the peak area of the sample and standard.

   α- Response factor

   w_1/w_2 is the weight or concentration of sample and standard.

   (b) **Calibration curve method:** Serial dilutions of standard solutions are prepared and injected in HPLC. A calibration graph is constructed by plotting peak area vs concentration. The concentration of the unknown sample is determined from the calibration graph.

   (c) **Internal standard method:** A known concentration of the internal standard (a compound with similar retention characteristics) is added to the standard solution and sample solution. The unknown solution concentration is determined using the ratios of the sample with internal standard and standard and internal standard. This method avoids the extraction steps when the sample matrix is complex in nature.

5. **Multi component analysis:** HPLC is also used for the determination of multiple components in a formulation simultaneously.

6. Isolation and identification of drugs and metabolites in biological fluids

7. Used in analyzing the mixture of components in natural products.

8. Useful in Biopharmaceutical and pharmacokinetic studies.

9. **Stability studies of drugs are performed using HPLC.**

10. Purification on the preparative scale can be carried out with HPLC.

CHAPTER 13

Ion Exchange Chromatography

Ion exchange is defined as a reversible reaction in which a mixture of similarly charged ions is exchanged using ion exchange resins. Cations or anions can be separated by this technique. The exchange of ions takes place based on the relative affinities of the ions for the ions in the ion exchanger. The ion-exchange chromatography is reported back from the 1850s.

PRINCIPLE

The basic principle is the attraction between oppositely charged particles. It is the reversible electrostatic interaction of charged species with the separation matrix. Ion exchange chromatography makes use of charge-charge interactions between the analyte in the sample solution and the functional groups which are present in the stationary phase (resin). The analytes are retained by the ionic interaction. The stationary phase surface presents ionic functional groups which interact with analytes of opposite charge.

The principle of separation is based on the reversible exchange of ions between the analyte ions in solution and ions in the ion exchange resin.

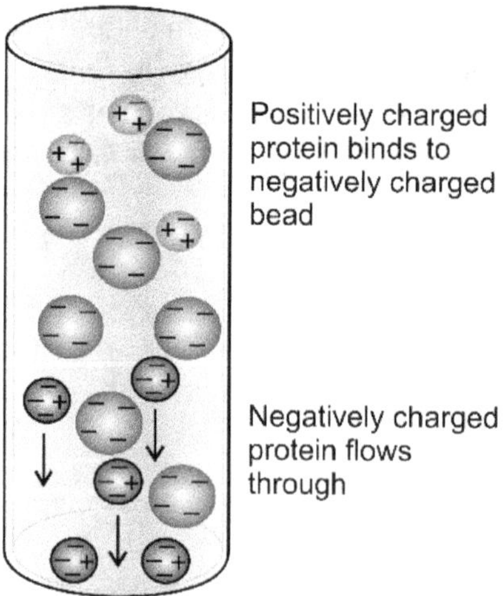

Fig. 13.1 Separation in Ion exchange chromatography.

TYPES OF ION EXCHANGE CHROMATOGRAPHY

(i)   Anion exchange chromatography

(ii) Cation exchange chromatography

In Anion exchange chromatography, anion exchange resins are used (positively charged resin), which binds to the negatively charged analytes (anions). E.g., Q anion exchanger, DEAE (Diethylaminoethylcellulose) – anion exchanger.

$$\text{Stationary-OH}^- + A^- \rightarrow \text{Stationary-A}^- + OH^-$$

In cation exchange chromatography, the cation exchange resins are used (negatively charged) and bind to positively charged ions (cations). They are used for the separation of cations in solution. E.g., Sulfonic acid – S-cation exchanger,
Carboxymethyl – CM cation exchanger.

$$\text{Stationary-H}^+ + M^+ \rightarrow \text{Stationary-M}^+ + H^+$$

The bound analytes are then eluted by changing the pH of the buffer or by increasing ionic strength.

Ion exchange resin should have the following requirements

1. It must be chemically stable, rigid and inert
2. It should be insoluble in common solvents
3. It should have a sufficient degree of cross-linking
4. The swollen resin must be denser than water

CLASSIFICATION OF ION EXCHANGE RESINS

Classification of Ion exchange resins based on the chemical nature

1. Strong cation exchange resin
2. Weak cation exchange resin
3. Strong anion exchange resin
4. Weak anion exchange resin

Table 13.1 Types of ion exchangers with examples.

S.No.	Type of exchanger	Functional group	Commercially available media
1.	Strong cation-exchangers	Sulfonic acid (S)	RESOURCE™ S, Macro-Prep® High S, Macro-Prep® 25 S
2.	Weak cation-exchangers	Carboxymethyl (CM)	CM sephadex C-50, C-25, Macro-Prep® CM
3.	Strong anion-exchangers	Quaternary Ammonium (Q)	RESOURCE™ Q, Macro-Prep® High Q, Macro-Prep® 25 Q
4.	Weak anion-exchangers	Diethylaminoethyl (DEAE)	DEAE sephacel A-50, DEAE cellulose, Macro-Prep® DEAE

Classification Ion exchange resins based on the source

Natural:

Cation: Zeolytes,clay etc

Anion: Dolomite

Synthetic: Inorganic and organic resins

Organic resins are the commonly used resins. They are polymeric resin matrix with ion-exchange sites. The exchange sites are present in the polystyrene of the resin, and di-vinyl benzene is the cross-linking agent which offers mechanical strength to the resin.

Structure of Styrene and Di-vinyl benzene

Styrene   Di-vinyl benzene

STRUCTURAL TYPES OF ION EXCHANGE RESINS

1. **Pellicular type:** The Particle size ranges from 30-40μ with an ion exchange film of 1-2μ thickness. **Their ion exchange capacity is very low.**

2. **Porous resin with coated exchanger beads:** Their particle size is 5-10 μ.They are porous in nature, and their efficiency is more

3. **Macro reticular resin bead:** A reticular network of the resin is present on the surface of the bead. It has low exchange capacity

4. **Bonded phase silica:** The surface of the support material is sulphonated and electro statically bonded to anion exchanger.

   It is **less efficient and has a low exchange capacity.**

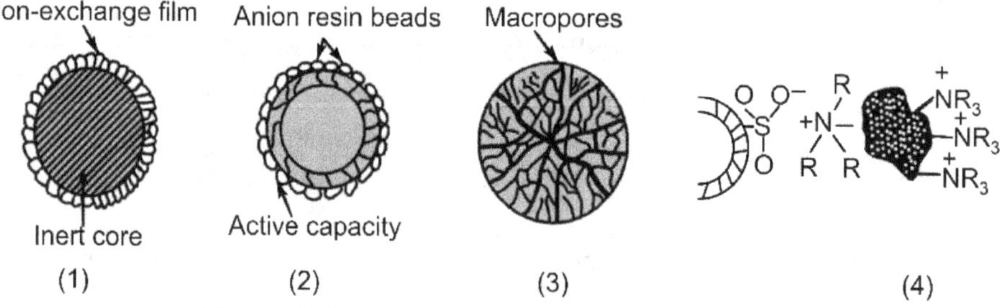

Fig. 13.2 Structural types of ion exchange resins.

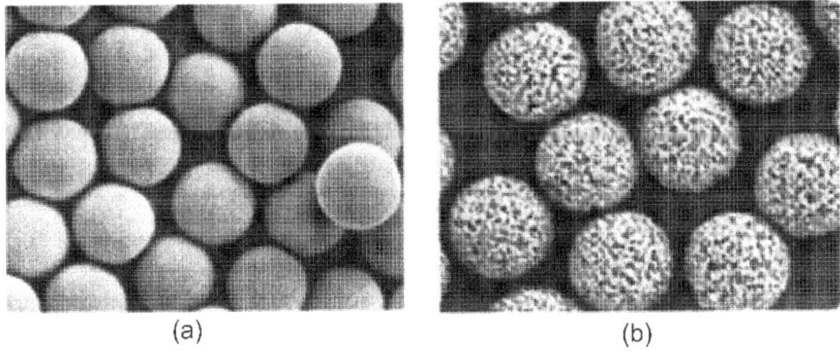

(a)   (b)

Fig. 13.3 Types of beads (a) Non-porous beads (b) Porous beads.

PROPERTIES OF RESINS

(i) **Particle size and porosity:** Decreased particle size offers more surface area, which in turn increases the rate of ion exchange. The range of particle size is 50-200mesh.

(ii) **Degree of cross-linking:** Increased concentration of cross-linking agents makes the resin rigid, leading to a decrease in swelling and decreased efficiency. On the other hand, the resin is soft, and swelling increases efficiency if the cross-linking agent concentration is less.

The quantity of cross-linking agents in the resin affects the separation.

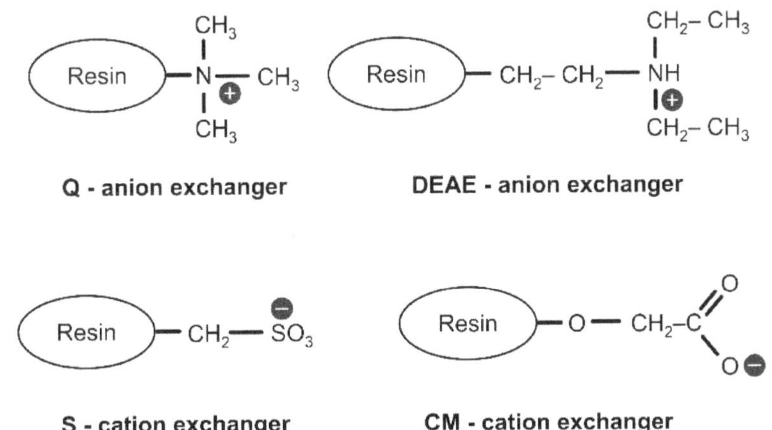

Fig. 13.4 Chemical structures of common ion-exchange media.

Mechanism of Ion exchange chromatography

The mechanism consists of 5 steps:

1. The ion diffuses to the exchanger surface. This occurs very quickly in homogeneous solutions.
2. The ion diffuses from the matrix surface to the exchange site, which is based on the degree of cross-linkage of resin and concentration of the solution. This step is considered to be the rate-determining step of the whole ion exchange process.

3. The exchange of ions occurs at the exchange site, which occurs instantaneously until equilibrium is attained.
4. The exchanged ion diffuses through the exchanger to the surface.
5. The molecules are separated by altering the pH or ionic concentrations or affinity elution, where there is greater affinity for the exchanger when the bound molecule is tightly bound to the stationary phase.

The ion exchange process is based on the affinity of the ions for the functional groups at the exchange site.

METHODOLOGY OF ION EXCHANGE CHROMATOGRAPHY

1. **Column and column packing:** Glass columns of 20:1,100:1 dimensions (length: diameter ratio) are used. Stainless steel columns are also used.

2. **Choice of ion exchange resin is based on**

   (a) Nature of ions to be separated-weak/strong

   (b) Charge on the ions to be separated –cation /anion.

   The efficiency of the ion is measured by ion exchange capacity. It is the total ion exchange capacity in terms of the exchangeable functional groups expressed as milli equivalents /gm of the resin. It is calculated for a cation exchanger as follows.

   A solution of sodium chloride of known concentration is passed through a column with the known weight of the resin. The acid eluting from the column is collected and titrated with a standard solution of alkali. The formula is

   $$\textbf{Capacity formula} = \frac{\textbf{V} \times \textbf{N}}{\textbf{W}}$$

   Where, V – volume of alkali solution,

           N – Strength of alkali solution in milli equivalents per litre,

           W – Weight of the resin in g.

   Similarly, anion exchanger capacity is determined by passing a solution of a definite concentration of sodium nitrate through the column. The concentration of eluted anions is determined by titration with a standard solution of silver nitrate.

3. **Column packing:** The wet packing method is used to pack the column. Resin is mixed with the mobile phase and packed uniformly in the column. The sample is dissolved in the mobile phase and introduced to the top of the column.

4. **Mobile phase:** The eluting solvents used are buffers, acids and alkalis of different strengths. E.g.,0.1N HCL, 0.1N NaOH, Phosphate buffer and acetate buffer, borate buffers, phthalate buffers etc.

   Organic solvents are not much useful. The pH should not be too low or too high to avoid interference with elution and high ionic strength is required for elution. pH should be 0.5-1 unit below pI (iso-electric point) in the cation exchanger and 0.5 -1 unit above pI in the anion exchanger.

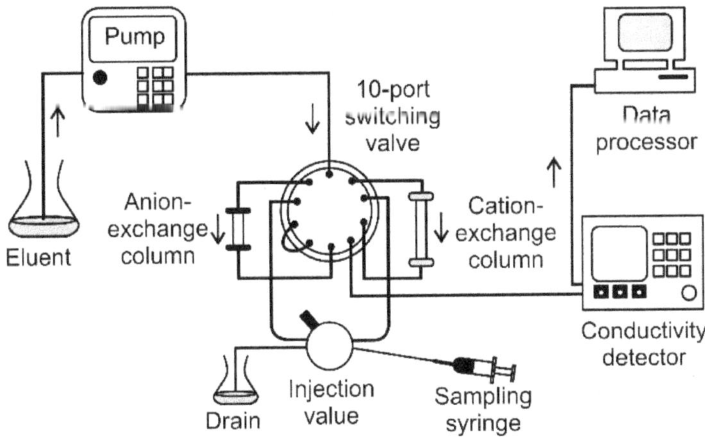

Fig. 13.5 Schematic diagram of Ion exchange chromatography.

5. **Sample preparation and loading:** The sample should be diluted, and it should be free of salt. It should be prepared with the same buffer which is used for the equilibration of the column. It should be filtered, and the sample size should not exceed the binding ability of the resin.

6. **Elution:** Components of the mixture separate and move down the column based on the affinity of the ion for the ion exchanger.

   There are two elution techniques **Isocratic** and **Gradient** elution

   Isocratic elution: The same composition of the solvent is used throughout the separation.

   Gradient elution: Increasing acidity or basicity of the mobile phase is used to separate complex mixtures.

7. **Detection:** Different fractions are collected based on time or volume and analysed by spectrophotometric, polarographic, conductometry or amperometric methods.

8. **Regeneration of ion exchange resin:** After separation, the ion exchange resin may not be useful for the next separation as exchangeable functional groups are lost.

But due to the cost of the ion exchange resins, they cannot be disposed off and hence reactivation is required.

REGENERATION

Regeneration makes the used ion exchange resin to be as efficient as a new resin. After the completion of the separation, the ion exchange resin should be regenerated by replacing the exchangeable cations or anions present in the original resin. Regeneration of ions is done as given below

Cation exchange resin is regenerated by charging the column with a strong acid like HCl.

Anion exchange resin is regenerated by charging the column with a strong alkali like sodium hydroxide or potassium hydroxide.

FACTORS AFFECTING ION EXCHANGE CHROMATOGRAPHY

I. Cross-linking and swelling

II. Nature of ion exchange resin

I. **Cross-linking and swelling:** Increased quantity of cross-linking agent makes the resin rigid, which leads to a decrease in swelling and decreased efficiency. On the other hand, the resin is soft, and swelling is more with increased efficiency if the cross-linking agent quantity is less.

Thus the quantity of cross-linking agents in the resin affects the separation.

Table 13.2 Weak and strong types of anion and cation exchangers.

Exchange Type	Ion exchange group	Buffer counter ions	pH range	Commercial samples
Strong cation	Sulfonic acid (SP)	Na^+, H^+, Li^+	4-13	Capto*S
				SP Sepharose*
				SP Sephadex*
				TSKgel SP_5PW
Weak cation	Carboxylic acid	Na^+, H^+, Li^+	6-10	CM Cellulose
				CM Sepharose*
				CM Sephadex*
				CM Sepharose* CL6B
				TSKgel CM-5PW
Strong anion	Quaternary amine (Q)	Cl^-, $HCOO_3$, CH_3COO^-, SO_4^{2-}	2-12	Q Sepharose*
				Capto*Q
				Dowex*1X2
				Amberlite*/Amberjet*
				QAE Sepharose*
Weak anion	Primary amine Secondary amine Tertiary amine (DEAE)	Cl^-, $HCOO_3$ CH_3COO^-, SO_4^{2-}	2-9	DEAE-Sepharose*
				Capto* DEAE
				DEAE Cellulose

II. **Nature of exchanging ions**

(i) **Valency of the ions:** The ion exchange **rate increases** with an **increase in valency** at low concentration and room temperature.

$$Na^+ < Ca_2^+ < Al_3^+ < Th_4^+$$

(ii) **Size of the ions:** Ion exchange rate **increases with a decrease in the size** of the hydrated ions

$$Li^+ < H^+ < Na^+ < NH_4^+ < K^+ < Rb^+ < Cs^+$$

(iii) **Polarizability: More polarisable** ions are preferred for ion exchange

$$I^- < Br^- < Cl^- < F^-$$

(iv) **Concentration: Dilute solutions favor** exchange of polyvalent anions

(v) **Charge on ions:** If the charge on the resin is higher than the charge on the solution, the exchange is favoured at a high concentration.

If the charge on the resin is lower than the charge on the solution, then the exchange is favoured at a low concentration

III. **pH of the mobile phase (buffer):** The pH should be one unit below pI for cation exchangers and one unit above for anion exchangers.

Cationic buffers are used in an anionic exchanger

E.g., Alkylamines, Tris, Aminoethyl alcohol

Anionic buffers are used in the cationic exchanger

E.g., Phosphate and acetate.

INSTRUMENTATION

General components of anion-exchange chromatography are given below

1. A high-pressure pump equipped with a pressure and flow indicator to deliver the eluent.
2. An injector for introducing the sample into the eluent stream and onto the column.
3. A column separates the sample mixture into individual components. The column can have an optional oven.
4. A detector to measure the analyte peaks as eluent emerges from the column
5. A data system for collecting and organising chromatograms and data.

APPLICATIONS

1. Water softening- Used for the removal of monovalent and divalent ions like potassium, calcium and magnesium
2. Demineralisation or deionisation of water - Removal of ions from water
3. Purification of solutions by removing ionic impurities
4. Separation of inorganic cations and anions
5. Used for the separation and purification of aminoacids, proteins, antibiotics, vitamins, fatty acids etc. in the pharmaceutical industry.
6. Isolation of dyes or metabolites from urine and blood
7. Used for the determination of the concentration of ionic solutions.
8. Used for the purification of proteins, nucleotides and small molecules
9. Separation of additives in food and drug samples.

CHAPTER 14

Gel Chromatography

Gel chromatography is a type of chromatography in which the separation of the molecules is based on **molecular size.** Gel chromatography is called Gel filtration chromatography, when the proteins and soluble polymers are separated using an **aqueous mobile phase.** When an **organic mobile phase** is used to separate **polar and organic soluble polymers,** the technique is called **Gel permeation chromatography. The gel chromatography is also called Size exclusion chromatography, Molecular exclusion chromatography, Molecular sieve chromatography.**

This technique uses porous particles as stationary phase and buffers as mobile phase. The stationary phase is a gel or a porous matrix with cross-linking in the form of beads.

PRINCIPLE AND MECHANISM

A mixture of compounds to be separated with different molecular sizes is passed through a porous matrix of stationary phase with buffer as mobile phase. The separation of the molecules is achieved on the basis of their molecular size and shape. The porous resins make use of their molecular sieving property to separate the components. The column packed with porous beads is equilibrated with the mobile phase (buffer), which fills the pores of the matrix and the interstitial spaces. Sometimes, the liquid inside the pores is referred to as the stationary phase, and this liquid is in equilibrium with the liquid outside the particles, referred to as the mobile phase.

Large molecular weight molecules are excluded from the pores, pass through the void space, interstitial spaces between the gel particles and they get eluted first. Smaller molecular weight particles pass through the porous matrix, enter into the pores of the gel and hence pass through the column at a slower rate, getting eluted later. Molecules with partial access to pores elute in the order of decreasing molecular size. Thus, larger molecules elute first, and smaller molecules elute later. Samples are eluted isocratically. (i.e.) There is no need of a different buffer for elution.

The distribution of molecules in gel filtration chromatography varies between 0 to 1. For larger molecules, K=0, which has no access to the mobile phase in the pores. For small molecules, K=1 can enter into the mobile phase in the pores. This variation of K between 0 and 1 helps in the separation of molecules based on their sizes, even with narrow molecular size variation.

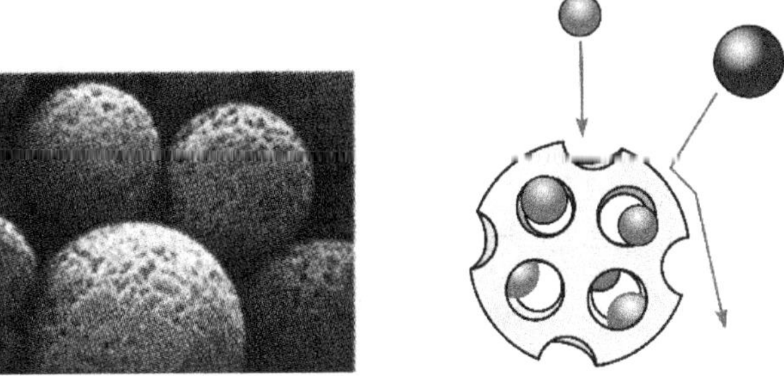

Fig. 14.1 Resin bead.

The resin bead is schematically represented as the ball. Large molecules cannot enter into the pore and travel with the mobile phase to get eluted quickly. But the molecules, which are smaller in size, enter in the pores and get eluted later.

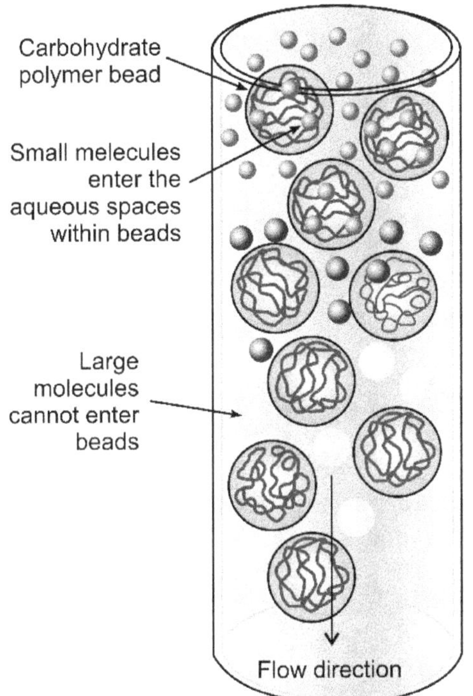

Fig. 14.2 Separation mechanism in Gel chromatography.

The eluted components can be detected by a suitable detector.

V_t = total column volume = $V_g + V_i + V_0$

V_g – Volume occupied by gel

V_i – Volume of the solvent in pores,

V_0 – Free solvent in column (space outside packing material)

V_e – elution volume of the compound

$V_e = V_0 + k_d V_i$

where k_d is distribution coefficient

V_c = Total (geometric) volume of the column

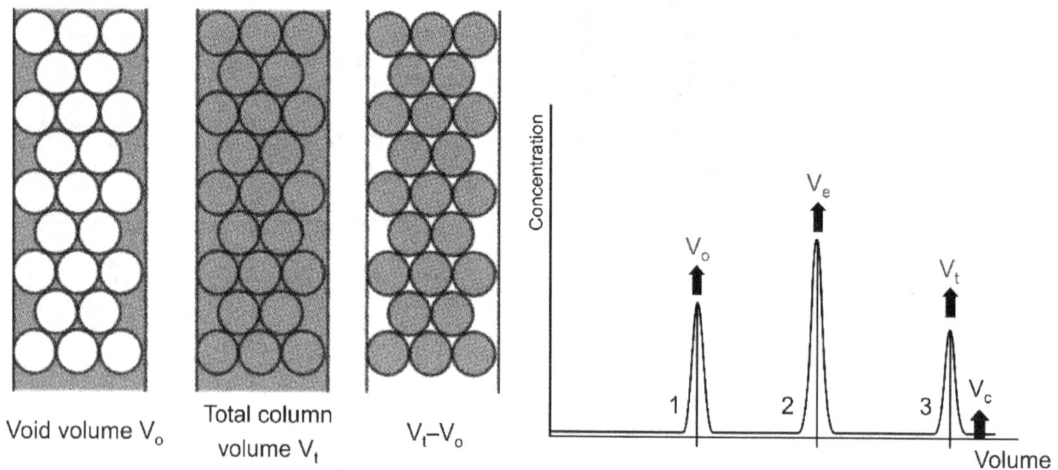

Fig. 14.3 Different volume terms used in Gel chromatography.

INSTRUMENTATION

The components of Gel chromatography are

1. Stationary phase
2. Mobile phase
3. Column
4. Pumps
5. Detectors

1. **Stationary phases:** These are semi-permeable, porous beads with a well-defined range of pore sizes. Cross-linked polymers are used as beads. The degree of cross-linking determines the pore size. Common stationary phases used are dextran, agarose or polyacrylamide. Some of the commercially available gel polymers are given below in table 14.1

Table 14.1 Some of the commercially available gel polymers.

Polymer gel	The molecular weight of molecules separated
Sephadex G (dextran based)	3000-70,000
	>70,000 m.wt excluded from the column
Bio gel P150 (Polyacrylamide based)	15,000-1,50,000 m.wt
Sepharose 6B (agarose based)	10,000-4,00,000 m.wt

Smaller pore sizes are used for rapid desalting of proteins or protein purification

Intermediate pore sizes are used for small protein separation

Very large pore sizes are used for the separation of biological complexes.

Dextran: Dextran is a homo polysaccharide of glucose residues. It is available as dry beads, and the beads swell when water is added. The trade name is Sephadex. It is used to separate small peptides and globular proteins with small to average molecular mass

Polyacrylamide: Cross-linking of acrylamide with N, N-methylene bis acrylamide gives these gels. They are used for the separation of small peptides and globular proteins. They are sold as bio-gel P. They are available in a wide range of pore sizes.

Agarose: Agarose is a Linear polymer of D-galactose and 3, 6 anhydro-1-galactose. The gel is held together with hydrogen bonds. The concentration in the gel determines the pore size. The pores of agarose gel are larger when compared with Sephadex or bio-gel P. Hence used for the separation of large globular proteins or long linear molecules such as DNA

Choice of the stationary phase depends on the molecular size, the shape of the molecules, operating pressure and temperature, flow rate, the viscosity of the sample, pH and tolerance for detergents and salts.

2. **Mobile phases:** Buffers are used as a mobile phase in gel filtration chromatography. E.g., phosphate buffer pH 7, Sodium chloride solutions, ammonium acetate, ammonium bicarbonate etc. Tris – HCl can also be used as a buffer.

   The sample is dissolved in the mobile phase and introduced into the column. Choice of the mobile phase depends on the type of separation and nature of the component to be separated.

3. **Columns:** Resolution increases with column length. Generally, long columns with small diameters are used. Length of the column can be 25,30,50,60 cm to 7.5-8 mm in diameter. Sometimes diameters of 2-3 mm are also used. If the column diameter increases, loading capacity increases as the bed volume of the column increases.

   V_t – total column volume, also called bed volume

   V_0 – void volume is the space between gel particles. It is usually $1/3^{rd}$ of bed volume.

   The larger particles elute through void volume.

   Smaller molecules enter into the mobile phase present in the pores of the stationary phase and elute later.

   $V_t - V_0$ = volume occupied by the gel (V gel), including gel matrix

   V_s (stationary phase) = $V_t - V_0 - V_{gel}$

   Column packing: There are three steps in column packing

   (a) Swelling of gel

   (b) Pouring of the gel into the column

   (c) Equilibration of the column.

   (a) **Swelling of gel:** It is done by two methods.

      Weighed quantity of the dried gel powder is mixed with the eluent (solvent) and allowed to swell. It is kept till equilibrium is obtained.

      The gel slurry is heated in a water bath at about 100° C. After swelling, the slurry is cooled and packed in the column.

The gel and the entire buffer used in the chromatographic column have to be degassed to avoid air bubbles. This helps in uniform packing of column giving good resolution.

(b) Pouring of gel in the column: The glass column is mounted vertically in a stand, and the buffer is poured through the column outlet to remove any air bubbles in the column. Now the gel slurry is poured with the help of a glass rod to avoid the trapping of air bubbles. The gel is allowed to settle down, and the excess buffer is decanted.

(c) Equilibration of the column: The column equilibration is done with the mobile phase buffer with a slow flow rate. The column is equilibrated with 2-3 bed volumes of the buffer.

4. **Pumps:** The pumps used in the HPLC systems are used here. They are
   1. reciprocating piston pump,
   2. syringe pump
   3. constant pressure pump

5. **Detectors:** The detectors used in Gel filtration chromatography can be classified as

   I. Concentration sensitive detectors: These detectors are further classified as
   1. Bulk Property Detectors- e.g.-Refractive Index (RI) Detector
   2. Solute Property Detectors- e.g.-Ultraviolet (UV) Absorption Detector
   3. Evaporative Detectors- e.g., Evaporative Light Scattering Detector (ELSD)

   II. Molar mass sensitive detectors: These detectors are classified as
   1. **Light Scattering Detectors:** Low Angle Light Scattering (LALS) Detectors
      Multi-angle Light Scattering (MALS) detectors
   2. **Viscosity Detectors- Differential Viscometers:** Other detectors used in Gel chromatography are Flame Ionization Detector (FID), Mass Spectrometer or Fourier Transform Infrared (FTIR) Spectrometer.
   1. **UV–visible detectors:** It is a photometric detector. Substances that absorb UV light can be detected by this solute property detector. These are versatile detectors with the best sensitivity and linearity. They are cost-effective, popular and widely used. There are three types of UV detector
      (a) Fixed wavelength detectors
      (b) Variable wavelength detectors
      (c) Photodiode array detectors
      (a) Fixed Wavelength detector: Detector operates at 254 nm only because most of the compounds absorb at this wavelength
         Simplest and cheapest of the UV/VIS detectors
         Limited in flexibility
         Limited in types of compounds that can be monitored

(b) Variable Wavelength detector: In this detector, a single wavelength is monitored at any given time, but operates in the wavelength range 190-900 nm.

More expensive requires more advanced optics

More versatile, used for a broader range of compounds

More sensitive due to photomultiplier tube

(c) Photo Diode Array Detector: In this detector, the absorbance of solutes is measured simultaneously at several different wavelengths. Light from the emission source, such as a deuterium lamp, is collimated by an achromatic lens system. The total light passes through the sample cell and then to a holographic grating. In this way, all wavelengths of light emitted by the source are passing through the sample. The dispersed light from the grating is allowed to fall onto a diode array. The array contains many hundreds of diodes, and the output from each diode is processed by a computer and stored.

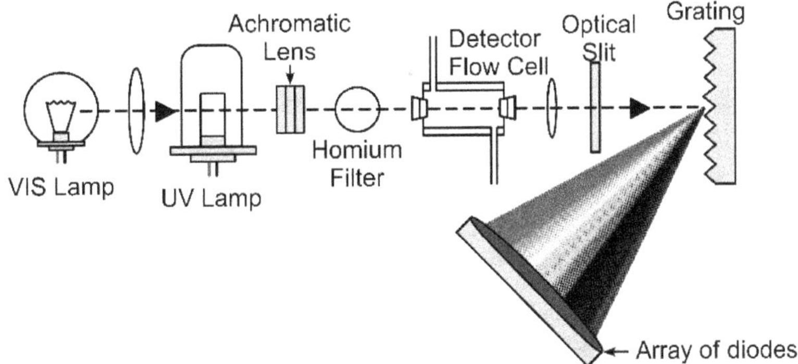

Fig. 14.4 Diode Array detector.

2. **Refractive index detector:** A Refractive index detector measures the molecule's ability to deflect light in a flowing mobile phase in a flow cell relative to a static mobile phase contained in a reference cell. The amount of deflection is proportional to the concentration of the solute in the mobile phase. The refractive index (RI) detector is one of the least sensitive LC detectors. This detector is extremely useful for detecting the compounds that are non-ionic, not showing ultraviolet light absorption and substances that do not exhibit fluorescence. E.g. sugar, alcohol, fatty acid and polymers.

3. **Fluorescence Detector:** These detectors are used for the substances that give fluorescence when subjected to radiation from a source. The excitation light is usually a low-pressure mercury lamp providing relatively high-intensity UV light at 253.7 nm. Many substances that fluoresce will be excited at this wavelength. The excitation light is focused by a quartz lens through the cell. A second lens, set normal to the incident light, focuses the fluorescent light onto a photocell. As it is very sensitive, its response is only linear over a relatively limited concentration range.

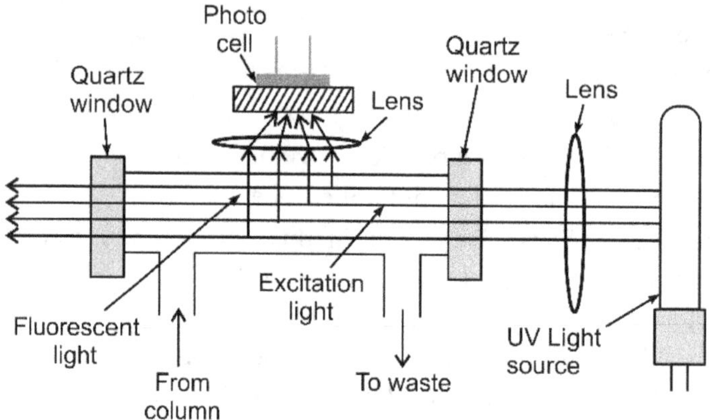

Fig. 14.5 Fluorescence detector.

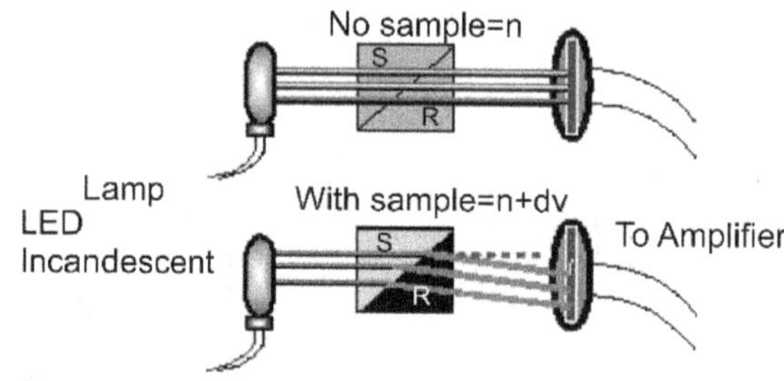

Fig. 14.6 Refractive index detector.

4. **Electrochemical detectors:** The current resulting from an oxidation/ reduction reaction of the analyte at a suitable electrode is measured in this detector. The current produced is directly proportional to the analyte concentration.

   The eluent should be electrically conductive.

   E.g. of drugs detected using this detector are phenol, catecholamines, nitrosamines, and organic acids in the ng range.

5. **Conductivity detectors:** This detector is used in ion-exchange chromatography for the detection of ionic compounds. The ability of the mobile phase to conduct current when placed in a flow cell between two electrodes is measured. Conductivity detectors measure the resistance, which is directly proportional to the concentration of ions present in the solution. Two electrodes placed in mobile phase form each corresponding to one arm of a Wheatstone Bridge.

6. **Evaporative light scattering detector (ELSD):** Detection is based on the scattering of a beam of light by particles of the compound.

   The three steps involved in detection are

       Nebulisation

       Desolvation

       Detection

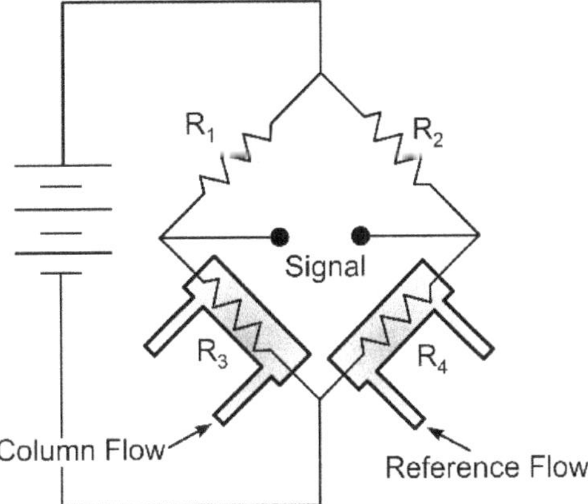

Fig. 14.7 Wheatstone Bridge.

A stream of inert gas nebulises the flow from the column. (Nebulizing is to convert a liquid into a fine spray or mist). The volatile mobile phase is evaporated, leaving tiny particles of the analytes. The particles, when passed through a laser beam, scatter the laser light. The scattered light is measured at right angles to the laser beam by a photodiode detector. It is a universal detector. The compound does not require a chromophore for detection by this detector.

7. **Mass spectrometer:** The HPLC system in which the Mass spectrophotometer is used as the detector is called LC-MS. In this method, the separation power of HPLC is combined with the detection power of Mass spectrometry. Mass spectrometry (MS) is a powerful analytical tool that can supply both structural information about compounds and quantitative data relating to mass.

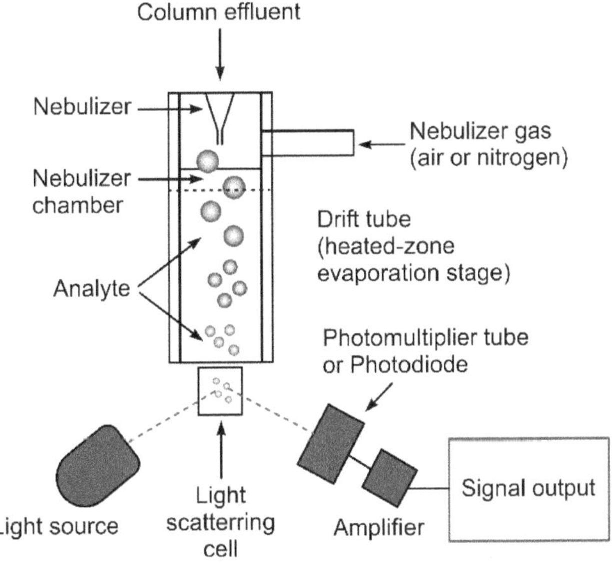

Fig. 14.8 Evaporative light scattering detector (ELSD).

8. **Chiral detectors:** Optically active compounds such as amino acids, sugars, terpenes and other compounds containing asymmetric carbon are detected by chiral detectors. There are two types of chiral detection techniques,
   1. Polarimetry or optical rotary dispersion (ORD)
   2. Circular dichroism (CD).

   ORD detectors are based on differences in refractive index

   CD detectors identify the enantiomers by measuring the differential absorption of right and left-handed circularly polarised light.

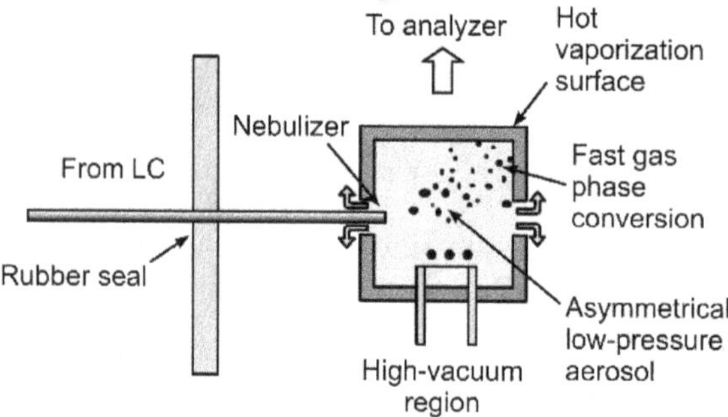

Fig. 14.9 Mass detector.

9. **Light Scattering Detectors:** An array of collimated detector elements is used to detect scattered light at discrete scattering angles.

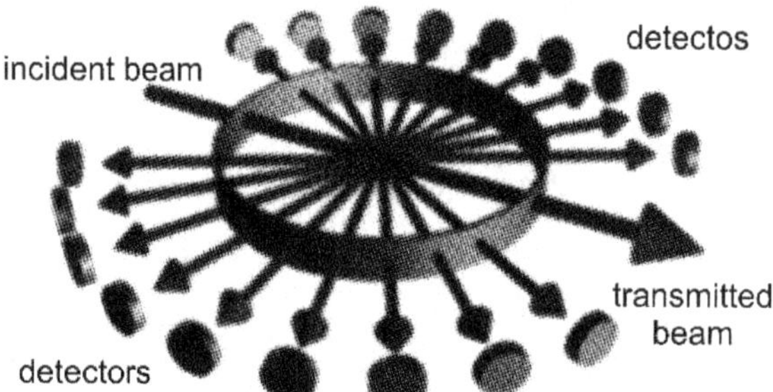

Fig. 14.10 Light Scattering devices.

Each sample is treated sequentially by a collimated light source, like a monochromatic laser, and the scattered light from each sample is measured as a function of scattering angle. This is shown schematically in figure 14.10, where an array of collimated diode detector elements detect scattered light at discrete scattering angles. This forms the basis of detection in these detectors.

A MALS detector employs three to twenty detectors arranged at various angles around the sample flow cell. A 90° detector is always included, and the remaining detectors are arranged as low as 12° and as high as 168°. A MALS detector measures the scattered light at a variety of angles, uses a model to extrapolate back to 0° to determine the molecular weight of the sample.

The RALS/LALS detector comprises of 90° and 7° detectors. For smaller samples of 10-15 nm in radius, the 90° detector is ideal because it is situated at a right angle to the incident beam, maximising the signal to noise ratio. The smaller samples scatter light with the same intensity in all directions. The RALS detector offers direct measurement of a sample's molecular weight. For samples with a radius larger than that 10-15 nm range displaying angular dependence, the LALS detector provides a measurement at 7°, which is the lowest detector position available.

Schematic diagram of Gel filtration chromatography instrumentation

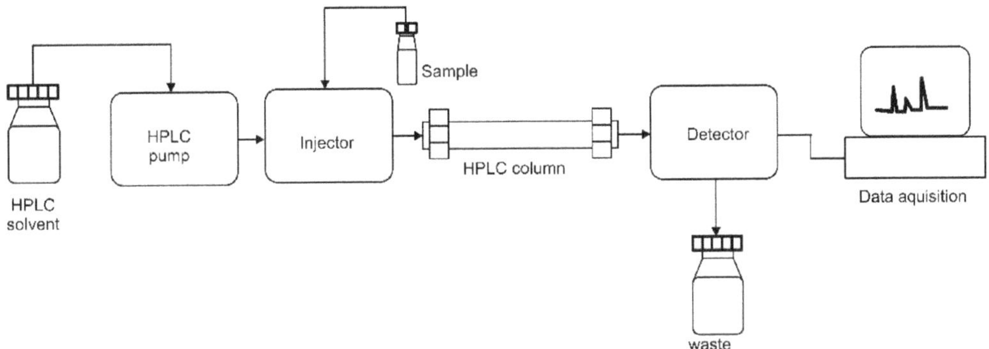

Fig. 14.11 Schematic diagram of Instrument in Gel filtration chromatography.

Working of gel filtration chromatography

1. **Preparation of column:** After packing the column as given above, it is equilibrated with the buffer.
2. **Loading of the sample:** The sample should be concentrated, and the sample size is normally 1% of bed volume. The sample should be filtered before loading. The sample is charged into the column, followed by the mobile phase.
3. **Elution:** Elution is carried out with the buffer, and the flow rate is maintained at 0.25 – 5ml/min. Fractions of individual components are collected separately. The same buffer is used for development as well as elution.

   V_e – elution volume is defined as the volume of the buffer that elutes from the column before a particular peak appears in the elution profile.

   K_d – distribution coefficient of the analyte

   $K_d = V_e\text{-}V_0/V_t\text{-}V_0$

   After elution, the column is washed thoroughly with 1-2 bed volumes of the buffer.
4. **Detection:** UV detector is the commonly used detector. To determine the molecular weight of the unknown compound, a calibration graph is drawn in semi-log paper. For calibration of the gel in the column, logarithms of the molecular weights of standard proteins are plotted against their respective ratios of elution volume to column void volume, the column void volume being the elution volume of a very large molecule such as Blue Dextran. The standard markers used are Yeast alcohol dehydrogenase (1,50,000 daltons). Bovine serum albumin (66,000 daltons), carbonic anhydrase (29,000 daltons), cytochrome C (12,400 daltons).

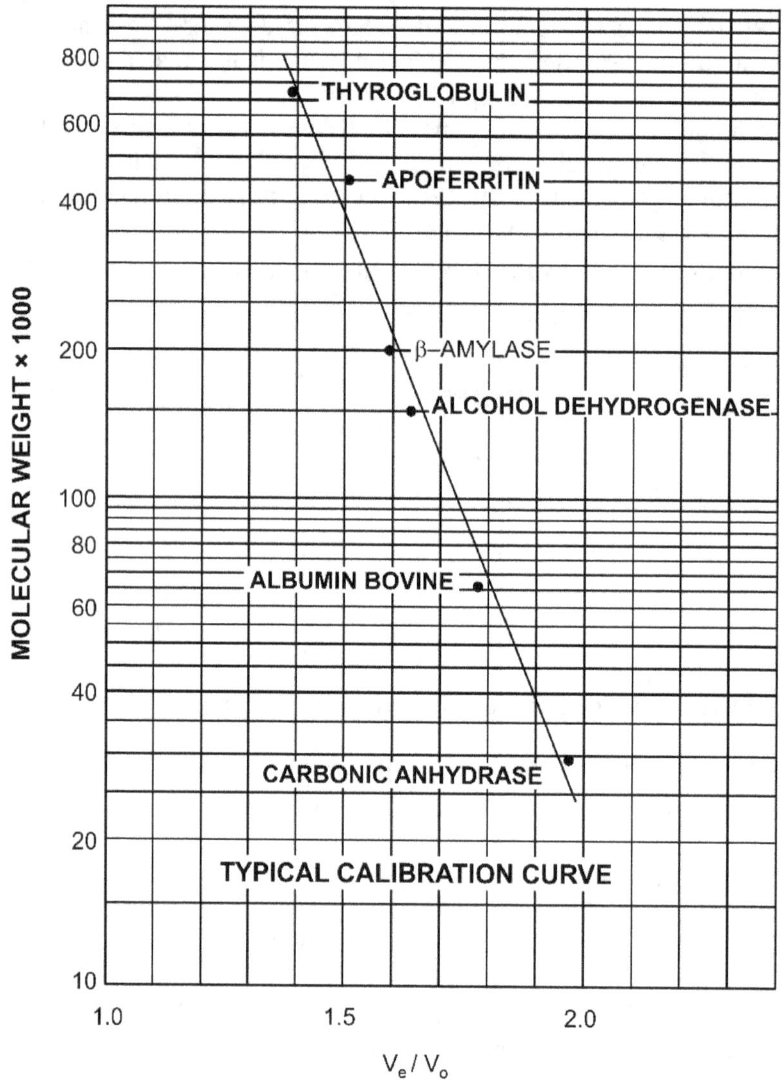

Fig. 14.12 Calibration graph used for molecular weight determination of unknown compound.

FACTORS AFFECTING RESOLUTION IN GEL FILTRATION

1. Column dimensions
2. Stationary phase packing density
3. Buffer composition
4. Sample volume and concentration
5. Particle size distribution
6. Flow rate of mobile phase
7. Pore size of packing
8. Viscosity of sample and buffer

Advantages

1. Biomolecules that are sensitive to changes in pH, the concentration of metal ions or cofactors and harsh environmental conditions can be analysed.
2. A significant advantage of gel filtration is that conditions can be varied to suit the type of sample without altering the separation.
3. Separations can be performed in the presence of essential ions or cofactors, detergents, urea, guanidine hydrochloride at a high or low ionic strength
4. Analysis can be performed at 37 °C or in the cold room according to the requirements of the experiment.
5. Molecules do not bind to the chromatography medium, so buffer composition does not directly affect resolution (the degree of separation between peaks).
6. Short analysis time and a small amount of mobile phase is sufficient.
7. Well defined separation with Narrow bands and good sensitivity.
8. There is no sample loss.

APPLICATIONS

1. **Molecular weight determination:** If the elution volume is known, the calibration graph can be used to determine molecular weight.
2. Separation of macromolecules like proteins and polypeptides, detection and separation of oligomers.
3. Purification of macromolecules, proteins, enzymes, amino acids, and polysaccharides. The impurities are separated based on their molecular weight.
4. Desalting- is the process of removal of salts from protein samples when salt is used in the separation of protein fraction. Salts travel slowly, whereas the large molecules get eluted faster in gel chromatography.

CHAPTER 15

Affinity Chromatography

Affinity chromatography is a sample purification technique, used mainly for biological molecules like proteins, enzymes etc. It is a type of chromatography that makes use of a specific affinity between the component to be isolated and a molecule that can bind to it (ligand).

PRINCIPLE

In this method, a mixture of proteins or nucleic acid molecules are separated by specific interactions with a compound known as ligand. Biological interactions between ligand and target molecule can be a result of electrostatic or hydrophobic interactions, Vander Waals' forces and hydrogen bonding.

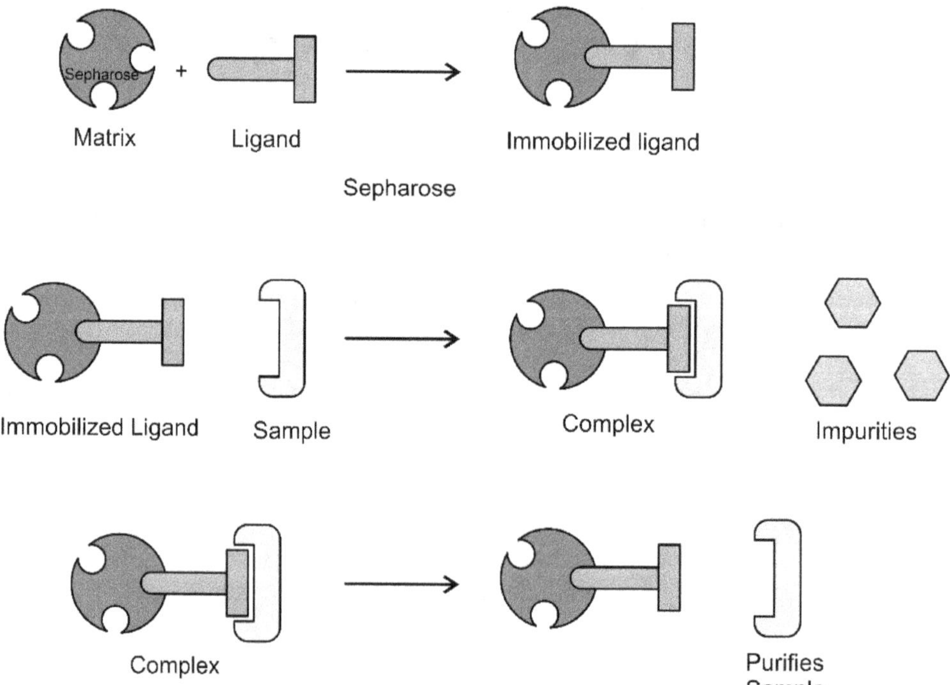

Fig. 15.1 Principle of Affinity chromatography.

The ligand is immobilized on a support. The interactions are specific biological interactions like enzyme-substrate, receptor-ligand or antigen-antibody reactions. These reactions are reversible reactions used for purification. The ligand which is attached to the solid matrix acts as the stationary phase. Ligand coupled matrices are commonly available

The molecule to be separated is dissolved in the suitable solvent and injected into an equilibrated column. The molecules with more affinity for the ligand are retained in the column and the substances with no affinity for the ligand are washed through the column.

The reversible binding of the ligand to the specific molecule or group of molecules form the basis of affinity chromatography

The elution of the retained molecules can be carried out by

1. Altering pH
2. Changing temperature
3. Changing ionic strength
4. Changing the organic solvent concentration

COMPONENTS OF AFFINITY CHROMATOGRAPHY

1. Matrix (support)
2. Spacer arm
3. Ligand

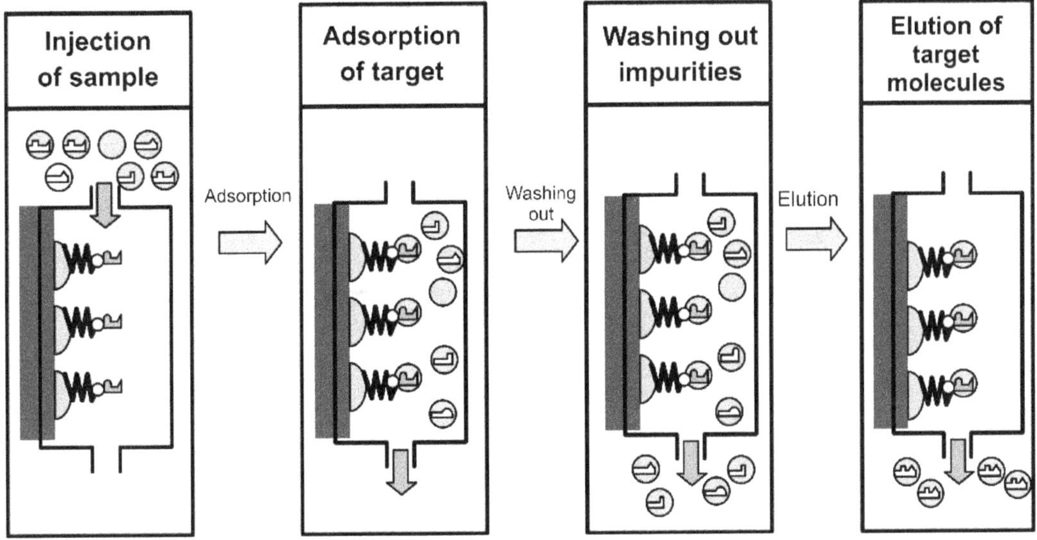

Fig. 15.2 Steps involved in Affinity chromatography.

1. **Matrix (support):** Matrix is the solid support to which the specific ligand is bonded covalently.

The characteristics of the matrix are

1. It should not be soluble in solvents and buffers used in the chromatographic development

2. It should not adsorb molecules to itself

3. It should be chemically and mechanically stable

4. The ligand or the spacer arm should be able to couple without undergoing any changes in its binding properties.

5. It should have a large surface area for the attachment of the ligand.

Agarose and polysaccharide are the most commonly used matrix materials. The other support materials used are Agarose (Sepharose), cross-linked dextran (Sephadex), Polyacrylamide, cross-linked cellulose, silica, methacrylate and polystyrene. They are usually used in the form of beads.

2. **Spacer arms:** Attachment of the ligand directly to the support matrix may alter the binding property of the ligand with macromolecules. To prevent this alteration, spacers or spacer arms are used.

The optimum length of the spacer arms can be 6-10 carbon atoms or equivalent to it. Spacers are used for small immobilized ligands. Eg of spacers :1, 6 diamino hexane and 6-amino hexanoic acid.

For agarose matrix, spacers like cyanogen bromide, 2-thiopyridyl, 1, 6 diamino hexane and 6-amino hexanoic acid can be used.

For polyacrylamide matrix, aminoethyl and hydrazide spacers can be used.

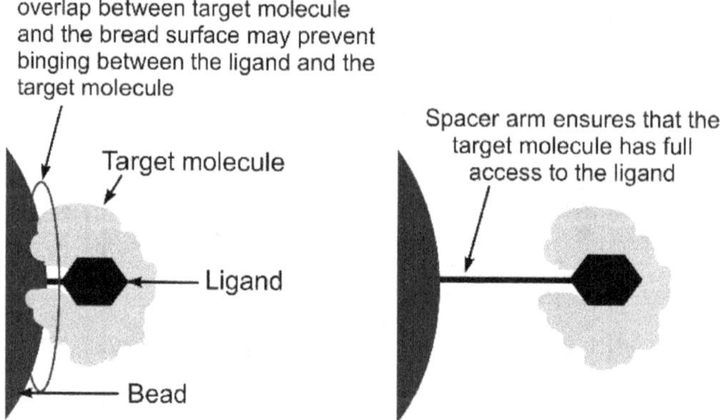

Fig. 15.3 Spacer arm.

3. **Ligand:** The ligand binds reversibly to the specific molecule or group of molecules and this forms the basis of affinity chromatography.

The choice of the ligand depends on

1. The binding affinity of the ligand and the target molecule.

2. The ligand should have a chemically modifiable group so that it can get attached to the support matrix without losing binding activity.

E.g., of ligands: coagulation factors
Heparin for lipoproteins, Proteins A and G for immunoglobulins
Antigen for antibody
Hormone for the purification of hormone receptor protein
Substrate for enzyme

SAMPLE PREPARATION

Samples should be clear and free from particulate matter. Sample preparation techniques including clarification are used to remove the components that interfere with binding (the interaction between the target molecule and the ligand).

The column must be pre-equilibrated in the binding buffer before sample application. Samples are applied at a high flow rate if there is a strong affinity between the ligand and the target molecule that quickly reach equilibrium. However, for interactions with weak affinity or slow equilibrium, a lower flow rate should be used.

PROCEDURE FOR AFFINITY CHROMATOGRAPHY

1. The affinity stationary phase is equilibrated with binding buffer
2. The sample is applied under the conditions required for the specific binding of the target molecule to the ligand. Reversible binding of the target molecules occurs, and the unbound material is washed through the column.

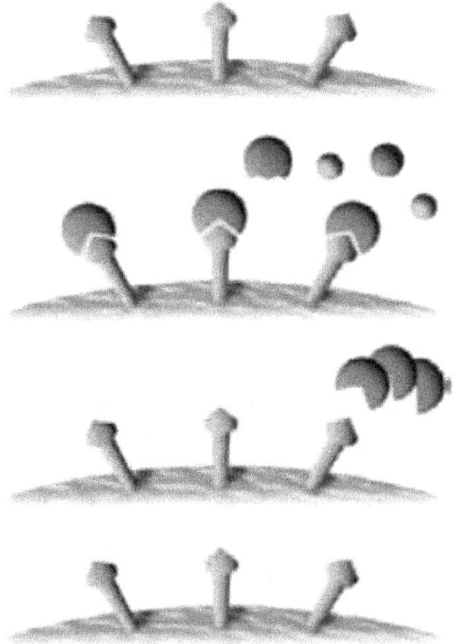

Fig. 15.4 Diagrammatic representation of procedure in affinity chromatography.

3. Elution of the bound protein is carried out by changing the working conditions. Elution can be performed by three methods
   (a) Specific method – competitive binding
      This method is carried out by using a substance that has more binding affinity for the ligand than the target compounds
   (b) Non –specific methods
      1. Changing pH
      2. Changing ionic strength or polarity
      3. Adding free ligands

(a) **Specific method – competitive binding:** In this type of competitive binding, A ligand with more binding affinity either for the ligand or for the target compound is used. These free ligands, when added to the column with ligand-target compound bindings, compete for binding with ligand or target compounds depending on their affinity. The figure shows the addition of a competitive free ligand with binding affinity for the target compound.

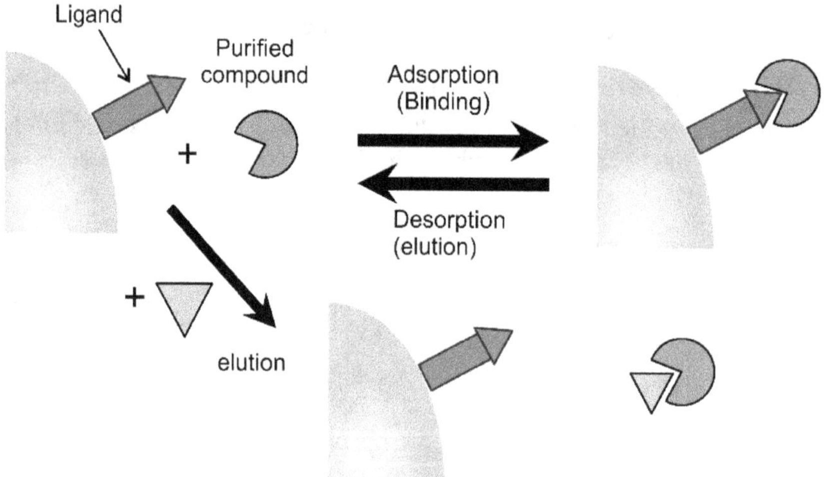

Fig. 15.5 Competitive elution of the target by adding a competitive, free ligand (triangle).

(b) **Non –specific methods**

pH elution: A change in pH alters the degree of ionization of charged groups on the ligand or the bound protein. This change may affect the binding sites directly, reducing their affinity, or cause indirect changes in affinity by alterations in conformation. A stepwise decrease in pH is the most common way to elute bound substances. The chemical stability of the matrix, ligand and target protein determines the limit of pH that may be used.

If low pH is used, the fractions are collected in neutralization buffer such as 1 M Tris-HCl, pH 9 (60–200 µl per ml eluted fraction) to return the fraction to a neutral pH. The column should also be re-equilibrated to neutral pH immediately.

Ionic strength elution: The exact mechanism of elution by changes in ionic strength depends on the specific interaction between the ligand and target protein. This is a mild elution using a buffer with more ionic strength, usually, NaCl is applied as a linear gradient. Enzymes usually elute at a concentration of 1 M NaCl or less.

Competitive elution: Selective eluents are used to separate substances when the binding affinity of the ligand/target protein interaction is relatively high. The eluting agent competes either for binding to the target protein or for binding to the ligand. Substances maybe eluted either by a concentration gradient of a single eluent or by pulse elution. In competitive elution, the concentration of the competing compound should be similar to the concentration of the coupled ligand. If the free competing compound shows weak binding than the ligand to the target molecule, a concentration of ten-fold higher than that of the ligand is used.

The reduced polarity of eluent: The polarity of the eluent is reduced to promote elution without inactivating the eluted substances. Dioxane 10% or ethylene glycol (up to 50%) are, e.g. of this type of eluent.

Chaotropic eluents: If other elution methods fail, deforming buffers, which alter the structure of proteins, can be used, e.g. chaotropic agents such as guanidine hydrochloride or urea. Chaotropes are likely to denature the eluted protein, and hence their usage is very limited.

4. **Re-equilibration of the column by the binding buffer.**

   The flow chart of the steps involved in affinity chromatography is given below.

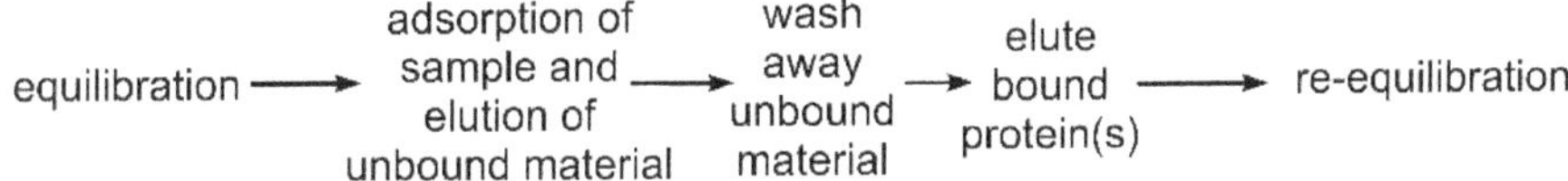

Advantages of affinity chromatography
1. Specificity is high and yields products of high purity.
2. Single-step purification and reproducible.
3. Matrix is reusable
4. The binding sites of the biological molecules can be studied.

Disadvantages of affinity chromatography
1. Ligands are expensive
2. Leakage of ligands in the mobile phase.
3. Solid support degradation causing limited lifespan
4. Possibility of non-specific adsorption

APPLICATIONS

1. Isolation and purification of biological macromolecules
2. Purification and concentration of enzymes
3. To study the binding of the compounds

4. To remove impurities in a mixture
5. Nucleic acid purification in genetic engineering
6. Antibody purification from blood serum in vaccine production.